DEVELOPMENT OF A DECISION SUPPORT SYSTEM
FOR GROUNDWATER POLLUTION ASSESSMENT

T0144476

VRIJE UNIVERSITEIT

Development of a Decision Support System for Groundwater Pollution Assessment

ACADEMISCH PROEFSCHRIFT
ter verkrijging van de graad van doctor aan
de Vrije Universiteit te Amsterdam,
op gezag van de rector magnificus
prof.dr. T. Sminia,
in het openbaar te verdedigen
ten overstaan van de promotiecommissie
van de faculteit der aardwetenschappen
op dindsdag 18 mei 1999 om 15.45 uur
in het hoofdgebouw van de universiteit,
De Boelelaan 1105

door
NEBOJŠA KUKURIĆ
geboren te Trebinje, Joegoslavië

A.A. BALKEMA / ROTTERDAM / BROOKFIELD / 1999

Promotoren:
prof. dr. I. Simmers
prof. dr. M.J. Hall

Published by
A.A. Balkema, P.O. Box 1675, 3000 BR Rotterdam, Netherlands
Fax: +31.10.4135947; E-mail: balkema@balkema.nl; Internet site: http://www.balkema.nl

A.A. Balkema Publishers, Old Post Road, Brookfield, VT 05036-9704, USA
Fax: 802.276.3837; E-mail: info@ashgate.com

ISBN 90 5410 418 X

CONTENTS

Abstract VII
Acknowledgements IX
Frequently used acronyms XI

1. Introduction 1

2. Encapsulation of knowledge for groundwater quality management

   2.1 Introduction 7
   2.2 Software review 8
   2.3 Groundwater quality management in the light of knowledge encapsulation 14
      2.3.1 Site characterisation 15
      2.3.2 Groundwater modelling 17
      2.3.3 Groundwater vulnerability assessment 20
      2.3.4 Groundwater remediation 22
   2.4 Artificial Intelligence 22
      2.4.1 Reasoning and knowledge representation 23
      2.4.2 Case-based reasoning 25
      2.4.3 Application of AI in groundwater quality management 28
   2.5 Concluding comments 30

3. Development of a Decision Support System (DSS): main considerations and framework

   3.1 Introduction 33
   3.2 A DSS as an integrated software system 34
   3.3 A DSS as an integrated information system 36
   3.4 A DSS for Groundwater Pollution Assessment - the framework 39
   3.5 The DSS knowledge component 41
   3.6 Summary 46

4. Knowledge-Based Module for Site Characterisation (SCM)

   4.1 Introduction 47
   4.2 Background 48
   4.3 Module content and organisation 51
      4.3.1 Task Unit ANALYSIS 52
      4.3.2 Other Task Units 61
      4.3.3 Discussion 64
   4.4 Software development and integration 66
      4.4.1 Hypertext and knowledge representation 67
      4.4.2 Other SCM features 71
      4.4.3 Adaptability 73

 4.4.4 Integration 74
4.5 Examples 75
 4.5.1 Hydraulic conductivity 75
 4.5.2 Solubility 79
Annex: glossary 82

5. Knowledge-Based Module for Vulnerability Assessment (VAM)

5.1 Introduction 85
5.2 Groundwater vulnerability in theory and practice 85
5.3 Description of methodology 88
 5.3.1 Likelihood of release and aquifer pollution 90
 5.3.2 Contaminant severity 93
 5.3.3 Pathway activity 94
 5.3.4 Target exposure 97
 5.3.5 Overall assessment score 98
 5.3.6 Discussion 98
 5.3.7 Testing aggregation procedure 99
5.4 Software development and integration 102
5.5 Case study 106

6. Knowledge-Based Module for Groundwater Pollution Modelling (GMM)

6.1 Introduction 111
6.2 Definition and ordering of modelling tasks 112
6.3 Knowledge acquisition 117
 6.3.1 Define Purpose 117
 6.3.2 Model Conceptualisation 120
 6.3.3 Modelcode Selection 129
 6.3.4 Model Design 133
 6.3.5 Calibration and Predictive Simulations 139
 6.3.6 Modelling Report 143
6.4 Knowledge systematisation and formalisation 149
6.5 Software development and integration 153
 6.5.1 GMM software 154
 6.5.2 REPORTER 159
 6.5.3 Modelling environment 161

7. Concluding remarks 165

References 169

Annex: Beyond engineering 179

Samenvatting 187

Curriculum Vitae 191

ABSTRACT

Computers have become the main tool used in groundwater management. Computer software has been developed for storage, processing and presentation of information on groundwater pollution problems. Continuing demands for more efficient handling of information have resulted in increasing integration of the software into Decision Support Systems (DSSs). Both quantitative (numerical) and qualitative (knowledge) information are required to support decision-making processes. To date, software development and integration were dedicated much more to numerical information than to knowledge. However, without knowledge components DSSs remain integrated software systems, whereas they should be primarily regarded as *integrated information systems*.

The DSS for Groundwater Pollution Assessment was conceptualised as an integrated information system. It consists, basically, of a DSS kernel, an application environment and a knowledge component. Design and development (and integration) of the knowledge component are the steps of the *knowledge encapsulation process* that is much more than transformation of knowledge in electronic form and its storage in the computer. The process begins with the definition of the problem and also includes (equally important) acquisition, systematisation and formalisation of knowledge. The case-specific nature and interdisciplinary character of groundwater problems are recognised as the main obstacles to effective encapsulation of knowledge in the field of groundwater management.

The *knowledge component* of the DSS for Groundwater Pollution Assessment is composed of task-oriented Knowledge-Based Modules (KBMs). Content and design of KBMs are strongly dependent on the tasks that they support, but in principle, they provide information on operations required for task accomplishment: which operations are required or recommended (e.g. acquisition, processing, presentation, etc.), why, how to perform them and when. So far, three KBMs have been designed and developed, namely: the Site Characterisation Module, the Vulnerability Assessment Module and the Groundwater Modelling Module. Various forms of knowledge representation, as offered by Artificial Intelligence, have been used in KBMs development. In addition, Case-Based Reasoning is anticipated as a promising approach to electronic encapsulation and processing of knowledge.

The booming development of Information and Communication Technology (ICT) opens new frontiers for knowledge encapsulation and DSS development in general. ICT techniques should be used to improve cooperation between experts as individuals, and between various fields of expertise (e.g. hydrogeology, geochemistry, ecology, software engineering, etc.). A sincere cooperation (teamwork in the real sense of the word) is equally crucial for establishment of a basic taxonomy of groundwater management tasks as for software integration; and for everything in between.

ACKNOWLEDGMENTS

My deepest gratitude goes to my brother dr. Ljubiša Kukurić who, in those years of war and despair, took care about everything that I could not, being so far away. I am no less thankful to Jaime M. Amezaga, for being a friend when I needed it at most.

Actual work on the thesis was done with the constant participation of my promotor prof. Michael J. Hall. He was always available when I needed him, and his experience, open-mindedness and pragmatism were of extreme importance for accomplishment of the research. Mike, many thanks.

Many thanks also go to dr. Yangxiao Zhou, my ex-mentor, ex-colleague and (still) my friend. He was 'the third' person constantly involved in the research. Therefore, I would compare the importance of his involvement with that of one of the angles for existence of a triangle.

The research was funded by the dr. ir. Cornelius Lely Stichting and the Rijksinstituut voor Integraal Zoetwaterbeheer en Afvalwater behandeling (RIZA). I would like to express my gratitude not only for the financing of the project, but also for the constant cooperation and support I received through my 'contact person' dr. ir. W.J. de Lange.

I am very grateful to prof. Ian Simmers for agreeing to share a role of promotor in the final stage of the research. I am also thankful to prof. J.Cunge, prof. R. Price and prof. J.J. de Vries, who read the thesis and provided me with very valuable comments.

Lastly (but only according to the course of events), I appreciate very much the support and understanding I obtained from my company, Netherlands Institute of Applied Geoscience TNO, being most often personified by ir. Jan Kooijman.

FREQUENTLY USED ACRONYMS

AI	- Artificial Inteligence
CBR	- Case-Based Reasoning
CDB	- Chemical Database
CS	- Contaminant Severity factor (in VAM)
DE1	- Dispersivity parameter module
DP	- Define Purpose (a Modelling Protocol step)
DSS	- Decision Support System
ES	- Expert System
GIS	- Geographic Information System
GMM	- KBM for Groundwater Pollution Modelling
(G)UI	- Graphical User interface
GWQM	- Groundwater Quality Management
GWL	- Groundwater Level
GWS	- Groundwater System
I(C)T	- Information (and Communication) Technology
KBS	- Knowledge-Based System
KBM	- Knowledge-Based Module
LI factor	- Likelihood of release and aquifer pollution factor (in VAM)
MC	- Model Calibration (a Modelling Protocol step)
MCO	- Model Complexity (in Modelling Protocol)
MD	- Model Design (a Modelling Protocol step)
MDO	- Management Decision Objectives (in MP)
MCM	- Model Complexity Module
MO	- Modelling Objectives (in Modelling Protocol)
MP	- Modelling Protocol in GMM
OAS	- Overall Assessment Score (in VAM)
PA factor	- Pathway Activity factor (in VAM)
RE1	- Retardation factor module
SC	- Site Characterisation
SCM	- KBM for Site Characterisation
TE factor	- Target Expsure factor (in VAM)
VA	- Vulnerability Assessment
VAM	- KBM for Vulnerability Assessment

1. INTRODUCTION

One of the first thoughts related to the possible (research on) development of a Decision Support System (DSS) was: 'it will not be about interfacing'. In other words, the research should not concentrate on connecting software into a DSS, because that is, in principle, a pure engineering task.[1] The research should be more about electronic Decision Support, and less about forming the operational System.

Decision Support Systems are the systems that support decision-making. In practice, the support just to be provided mostly through the electronic storage, processing (modelling) and visualisation of data (numerical information). Although a knowledge base was often mentioned (next to a model and a database) as a basic DSS component, very few DSSs were found that contained any kind of knowledge component. This is a notable result, because both data (*quantitative information*) and knowledge (*qualitative information*) are needed to support decision-making. Without a knowledge component DSSs remain integrated (interfaced) software systems, whereas they should be *integrated information systems*. This conclusion led to clarification of research objectives; the research should be (and has been) dedicated to a DSS knowledge component, or, in somewhat broader terms, to the 'electronic encapsulation of knowledge'.

Knowledge about what? In a time of 'interdisciplinary problems' none of the possible terms seemed to be exact enough. At the beginning of the research, knowledge about groundwater problems in general was targeted, with the intention of narrowing the scope during the research. Knowledge coming from various disciplines (hydrogeology, geochemistry, geophysics, ecology, economy, legislation, etc.) is involved in the management of groundwater problems. Therefore the first phase of the research was conducted under the umbrella 'electronic encapsulation of knowledge for Groundwater Quality Management' (GWQM).[2] It included:

– a review of current achievements in software integration and knowledge encapsulation;
– a review of postulates and techniques of Artificial Intelligence (AI) related to knowledge encapsulation and DSS development, and
– an analysis of Groundwater Quality Management in the light of knowledge encapsulation.

The purpose was to define a possible realm and the best way(s) of applying AI in GWQM. The

---

[1] Already at that time, interfacing seemed to became a 'common practice'. Four years later, it appears that the epithet 'fully integrated' can be assigned to a very few DSSs. This is despite all advances in software and hardware, and an immeasurable amount of work on interfacing done in the meantime. Interfacing is the engineering task, but - apparently - a tough one.

[2] The term 'Quality' was added to emphasise consideration of both quantitative and qualitative aspect of groundwater problems.

research phase was completed at the end of 1996 and results were published as an extended article in *Water Resources Management* (Kukurić and Hall, 1998). The article has been included in the thesis in its original form (Chapter 2), showing that no major conclusions needed to be reconsidered, this in spite of breath-taking developments in the field of AI and Information Technology (IT) in general.

The results obtained in the first research phase allowed the concept 'DSS - an integrated information system' to be worked out (Chapter 3). The concept and defined application realm were subsequently used to establish the framework of a DSS for Groundwater Pollution Assessment at the local scale (Chapter 3). The DSS consists of the DSS kernel (that is a fully integrated powerful rational data base and a GIS), an application environment and a knowledge component.[3] Since the knowledge component was the prime target of the research, Chapter 3 also contains general considerations on knowledge encapsulation, knowledge processing and the DSS knowledge component.

The second research phase was dedicated to the design and development of the DSS knowledge component. The component consists, at the time of writing, of three task-oriented Knowledge-Based Modules (KBMs), namely:

- a Site Characterisation Module (SCM);
- a Vulnerability Assessment Module (VAM), and
- a Groundwater Modelling Module (GMM).

Characterisation of an investigated site is the first task to be carried out when dealing with point-source groundwater pollution problems. The main objectives of the characterisation are conceptualisation of the groundwater system and diagnosis of groundwater pollution. General knowledge on site characterisation has been acquired, systematised and encapsulated in the Site Characterisation Module (Chapter 4). The SCM has been encoded in HyperText Markup Language (HTML) that allows encapsulation of large amounts of semi-structured knowledge in an easily adaptable form. Besides knowledge, the SCM contains the links with procedures for data storage, processing and presentation located elsewhere in the DSS. The module was presented at the international conference 'Hydroinformatics 98' (Copenhagen, Denmark) and the paper was published in the conference proceedings (Kukurić et al, 1998a).

In general terms, groundwater Vulnerability Assessment (VA) can be described as a procedure for the quick assessment of groundwater pollution potential. It is based on intrinsic aquifer characteristics, though contaminant characteristics and management practice can also be taken into account. In the context of investigations of groundwater pollution potential (at a local scale),

---

[3] The DSS kernel was provided by the REGIS package, a regional geohydrological information system developed by Netherlands Institute of Applied Geoscience TNO. Although some assistance was received from TNO (gratefully acknowledged) with interfacing REGIS and the knowledge components, the programming work associated with the latter was carried out entirely by the author.

VA is seen as a step that follows characterisation of the investigated site. Accordingly, VA uses the results of the site characterisation, yielding a first assessment of the pollution potential. The results of VA provide a basis for further investigations and/or assessment, and a means for comparison of pollution potentials. A new ranking-based VA methodology has been developed and encapsulated in the VAM-Vulnerability Assessment Module (Chapter 5). The VAM has been developed in object-orientated Borland Delphi Developer, a tool based on Object Pascal as a programming language. Integration of the VAM into the DSS involved (inter alia) DSS kernel, the SCM and a Chemical Database (CDB). The module was presented at the international conference 'Hydrology in a Changing Environment' (Exeter, United Kingdom) and the paper was published in the conference proceedings (Kukurić et al, 1998b).

The GMM has been developed to support modelling of point-source pollution problems (Chapter 6). It consists of several software applications developed in various programming environments. The core of the GMM is an electronic Modelling Protocol guidance composed of sets of hypertext-based topics. The knowledge on general model complexity is encapsulated in a rule-based Model Complexity Module by using a knowledge base editor. The GMM also contains two additional modules (developed in Delphi) that assist in estimation of the retardation factor and dispersivity. The Modelling Protocol acts as a platform that integrates the software applications in a unique DSS module. In the framework of GMM development, MODFLOW, a modular 3D groundwater flow model, has been fully integrated with the DSS kernel. The GMM was presented at the international conference 'MODFLOW 98' (Golden, Colorado) and the paper was published in the conference proceedings (Kukurić et al, 1998c).

Development of above-mentioned KBMs (described in chapters 4, 5 and 6) completed the second phase of the research. Concluding remarks about the conducted research are given in Chapter 7.

Although not being explicitly stated in the title of the thesis, the research described in the following chapters was primarily about knowledge encapsulation. The title reflects the purpose of knowledge encapsulation: to provide 'decision support' for 'groundwater pollution assessment'. Still, the emphasis was on conceptualisation and prototyping of a knowledge-containing DSS. Decisions are, in principle, based on available information, meaning that an electronic DSS should integrate both quantitative and qualitative information (data and knowledge), in order to provide adequate support to the decision-making process. Therefrom arises the concept of a DSS as an 'integrated information system'. Knowledge in the DSS is partly about numerical information, for example: which method should be used to define a parameter 'A' in the present situation, or which value of the parameter 'A' is appropriate for present situation? Additionally, knowledge allows the DSS to act as a process-oriented or a task-oriented system; in other words, the DSS can support decisions on steps that should be taken (e.g. with respect to data acquisition, processing, presentation, etc.) in order to carry out the task (to solve a posed problem).

Knowledge encapsulation in the field of groundwater quality management appears to be a real challenge, as thoroughly discussed in Sections 2.4 and 3.5. Until recently, AI has dealt exclusively with fairly structured knowledge coming from narrow, well-defined domains. Accordingly, AI techniques and procedures for knowledge acquisition and representation are developed to 'capture' this kind of knowledge. Numerous interviews have been conducted with the groundwater experts (as a part of this research), in an attempt to locate knowledge domains suitable for electronic encapsulation. Apparently, those domain exist and could be defined (e.g. expertise on validity of equilibrium sorption isotherm or on trade-off between network density and number of particles - see Section 2.4.3). That was, however, not the only conclusion that emerged from the conducted interviews; much more striking was the revelation of ambiguousness and rigidness of a 'global picture' in which narrow local domains play a relatively minor role. In other words, global approaches to groundwater pollution problems have not been worked out sufficiently (e.g. for Site Characterisation), or various approaches have been introduced and used (e.g. Vulnerability Assessment) without thorough comparison of their applicability in various situations. The same holds for various methods and tools that, once made available and implemented, are continuously used without reconsideration. Large portions of what is considered as common, general knowledge is not sufficiently structured, widely scattered among various sources of information (books, reports, guidelines, etc), often incomplete and sometimes even inconsistent. Processing knowledge within narrow domains can be successful only if input information coming from the (more) global domain is correct; e.g. what is the sense of adjusting network density and number of particles (in a groundwater model) if the modelling objective or the conceptual model are not set up properly? Moreover, the most important decisions (and the most serious errors) are made at a global level.

Gathering and ordering of general knowledge on groundwater pollution problems (as done in this research) is very labourious task, but is certainly worth doing. Some of the encapsulated knowledge might, in eyes of a field expert, look like no more than a set of 'standard receipts'. Not everyone is, however, expert in a particular field. Besides, encapsulated knowledge can always serve as a useful reminder. Currently, much knowledge is not used in daily practice simply because it is not available (in consistent, systematic form) at the moment when it is needed. Contemporary IT offers possibilities for systematic storage and transfer of large portions of electronically encapsulated information; once available in a computer (in an appropriate form), it will most certainly be used.

General knowledge is by definition explicit and its acquisition does not ask for either 'classical' interviews with experts or other common AI acquisition procedures; the need for an AI acquisition procedure becomes evident when entering deeper domains that contain much more heuristic, tacit knowledge. Only one deeper knowledge domain has been entered during the research: (groundwater) model complexity (Chapter 6). An extensive (electronic) questionnaire was prepared and sent to experts. Response to the questionnaire was very limited, reflecting a lack of motivation for knowledge sharing. Cooperation has also proved to be crucial for systematisation of general knowledge, this time having 'consensus' as its most important postulate (see Section 3.5). The general lack of unwillingness of experts to cooperate has triggered thoughts on ethical and philosophical aspects of knowledge sharing. It was very tempting to continue research purely in this direction, but eventually a sense of practicality prevailed.

It was no less tempting to enter deeper into the innovative and sometimes fascinating technological world of Artificial Intelligence, where knowledge-based agents cooperate without hesitation and where DSSs are, at least conceptually, really 'smart' systems. Again, this was not such a good idea, knowing how conservative the hydrogeological community is (generally speaking). Eventually, this research was about implementation of AI and IT in hydrogeology into an extent ('just one step from reality') that might be accepted (appreciated?) by hydrogeologists. This research was not about generating new hydrogeological knowledge and also not about developing new IT technologies. It was about bringing new technologies (and maybe a new way of thinking) to hydrogeology.

All the steps of the knowledge encapsulation process were carried out in order to create a hydrogeological integrated information system:

- major tasks related to groundwater pollution assessment were worked out in detail;
- general knowledge required for accomplishment of the tasks was acquired and systematised;
- AI knowledge representation forms were applied in formalisation of knowledge and design of knowledge-based modules,
- the modules were developed and integrated into the DSS using state-of-the-art IT techniques.

The developed (prototype) DSS for groundwater pollution assessment appears to be one of the first attempts in the field of hydrogeology to encapsulate and actively to integrate knowledge in a comprehensive task-oriented DSS. Its development will be justified if it inspires further electronic encapsulation and integration of information on groundwater problems.

At this point, the introduction can be considered completed because it has outlined (mainly by introducing the following chapters and some follow-ups) what this thesis is about (and not about). There are, however, still those 'tempting' aspects of knowledge that (with a few exceptions) have not been mentioned explicitly in the thesis, although they were continuously dealt with during the research. Therefore an annex to the thesis is added, containing some thoughts about technological, philosophical and ethical aspects of knowledge.

2. THE ELECTRONIC ENCAPSULATION OF KNOWLEDGE FOR GROUNDWATER QUALITY MANAGEMENT

2.1 Introduction

Decision-making is one of the key processes of human cognition. Decisions are based on available information, so that correctness of a decision, in principle, depends on information, its availability and quality. Methods have been developed to acquire, process and present information. Most of the methods for information processing and presentation are desk-top methods that have been encoded and electronically encapsulated into what are commonly known as software tools. During the last few decades, numerous software tools have been developed for e.g. modelling, statistical processing and graphical presentation of information. Beside software tools for processing and presentation, electronic data bases have been developed to accommodate (store) tremendous amount of information that has become available. Attempts have also been made to acquire, formalise and encode qualitative information (knowledge) into Expert Systems (ES) or Knowledge-Based Systems (KBS) that could support, or even replace, human decision making[1].

Quick decisions are an imperative of modern society, thereby posing a strong care for integration of information, i.e. for integration of software tools developed to store, process and present information. Integration of both quantitative and qualitative information is required to support (accelerate, as well as improve) decision-making processes. As a result of its needs, modern society has promoted the development of integrated software tools called Decision Support Systems (DSSs) or, since recently, more often, Intelligent DSSs (IDSSs).

GroundWater Quality Management (GWQM) has also been substantially influenced by software development. Encoded numerical groundwater models are the main tools used nowadays to assist in decision-making for groundwater pollution problems. Development of computer graphics has encouraged evolution and widespread application of vulnerability assessment methods. Data bases are extensively used in GWQM, playing often a multifunctional role that reaches far beyond 'simple' data storage. During the last decade, a strong tendency has become evident to integrate GWQM software tools in order to enhance and accelerate their communication. At the same time, courageous steps have been made towards encapsulation of the knowledge needed for GWQM. The next section of this paper contains a review of software developed to date that has contributed towards integration and encapsulation of information on groundwater pollution problems. The review is used as a basis for a discussion that extends over two subsequent sections. Firstly, some aspects of GWQM that are found important for knowledge encapsulation

---

[1] A term 'qualitative information' is here used as a synonym for knowledge, whereas the term 'quantitative information' denotes information that is numerical or that can be quantified.

and development of DSSs are developed and scrutinised. Subsequently, techniques of Artificial Intelligence (AI) that are (or might be) used for GWQM knowledge encapsulation and that influence design of DSSs are reviewed and discussed. The most important findings are summarised in the last section of the chapter.

2.2 Software review

Prior to reviewing the software that has been developed, one terminological clarification is necessary; it is related to the use of the terms Expert System (ES), Knowledge Base (KB), Knowledge Based System (KBS) and (Intelligent) Decision Support System ((I)DSS). Inconsistent use of these terms causes some inconvenience when reviewing the software, as well as while addressing the question of future development of knowledge-based (knowledge-containing) software tools for GWQM: what should they be: ESs, KBs or KBSs or (I)DSSs?

'Decision Support System' (DSS) and 'Intelligent Decision Support System' (IDSS) are the terms that are nowadays widely used in practice. The terms themselves are quite self-explanatory: the systems that support decision-making. 'System' implies 'organisation' that is needed when a number of components (of a system) are brought together, connected and simultaneously or subsequently used. 'Intelligent' suggests an IDSS capability to make inferences, and/or to communicate with a user in an 'intelligent' manner. Avoiding rigid definitions of expert systems, it is sufficient to state that the inferencing is the essential characteristic of ESs. The term 'Expert System' is nowadays almost completely replaced by the term 'Knowledge Base', emphasising a less pretentious role to provide expertise and the increased incorporation of other kinds of knowledge besides the heuristic. The term 'Knowledge-Based System' is even more in use indicating again a complex (multi-component) structure of a software tool. If a DSS contains any qualitative information (knowledge) it can be also called a KBS. This apparent transition from ESs to (I)DSSs, enabled by development of information technology, is manifested first of all by the integration of software tools for information processing and of (encapsulated) complementary knowledge. Although a terminological differentiation can still be imposed on the knowledge-based software tools (see e.g. Doukidis, 1988, for discussion), practice shows quite liberal (but not consistent) use of the terms. Therefore, the following review covers all (acquired and relevant) integrated and/or knowledge-containing software tools related to GWQM, regardless of the terminology used (Table 2.1). It is striking that none of more that twenty expert systems described in the UN document 'Expert Systems in Hydrology' (Bakonyi, 1993) is found relevant to be included in this review.

The first attempts to encapsulate knowledge on GWQM into ESs were evident more than a decade ago. ESs were, especially at that time, developed for narrow, well-defined knowledge domains, where (particularly heuristic) knowledge could be acquired and formalised in sets of rules. Such domains could not be clearly defined with respect to poorly-structured GWQM tasks (with exception of ranking procedure for waste disposal sites), so these attempts died out. At the same time, integration of hydrogeological software started, stimulated by the rapid development

Table 2.1 Software Overview

SOFTWARE & DEVELOPER	PURPOSE	INTEGRATED SOFTWARE COMPONENTS		
		Data Bases / GIS	Models / Methods	Embedded Knowledge
GWW, UNESCO 1995	storage, processing, presentation	Hg DB, Ch DB (to be filled in)	set of methods (mainly statistical)	
HGDB [Data Base] [a] Newell et al 1990	site characterisation & GW modelling input preparation	Hg DB (including extensive statistics)	hydrogeologic settings (DRASTIC)	
REGIS [Information System], TNO 1994	storage, processing, presentation	Hg DB and Ch DB, GIS	set of methods (mainly statistical)	
... Biesheuvel & Hemker 1993	simulation and prediction of GW flow	GIS	numerical GW flow model	
... Roasa et al 1993	sim. and pred. of GW flow and GW cont.	GIS	GW flow and cont. transport model	
Wellhead Modelling User Interface Rifai et al 1993	delineation of well head protection areas	GIS	integrated semi-analytical GW flow model	
... Paget 1994	GW vulnerability assessment	GIS	DRASTIC VA method	
... Evans and Mayers 1990	GW vulnerability assessment	GIS	DRASTIC VA method	
... Rundquist et al 1991	GW vulnerability assessment	GIS	DRASTIC VA method	
Integrated GW models Fedra 1993	GW quality management	GIS	GW flow and cont. transport model	integrated rule-based ES (interfacing assistance and consistency checking)
ROKEY [Expert System] McClymont and Schwartz 1987	simulation and prediction of GW contamination	modest Hg and Ch DB	GW flow and cont. transport model (analytical solution)	EXPAR rule-based (forms)
OASIS [DSS] Newell et al 1990	simulation and prediction of GW contamination	Hg DB, Ch DB	GW flow and cont. transport models	EXPAR (converted in hypercard system)

Table 2.1 Continued

SOFTWARE & DEVELOPER	PURPOSE	INTEGRATED SOFTWARE COMPONENTS		
		Data Bases / GIS	Models / Methods	Embedded Knowledge
KGM [Knowledge Based System] Saaltink and Carrera 1992	model conceptualisation and model code selection	model code DB		rule-based (decision tables) contains heuristic knowledge
MARS [Data Base] IGWMC 1992	model code selection	model code DB		rules for data-base operation
MODELEXPERT [ES] Sc.Soft.Group 1995	model code selection	model code DB		rules for data-base operation
GEOTOX [ES] Wilson et al 1987	assessment of waste disp. sites		ranking method	rule based (semantic network)
Pattern Recognition System [ES] Datta et al 1989	identification of GW pollution source	concentrations patterns DB	statistical pattern recognition method and ranking method	rules for selection of source location and magnitude
EXPREX [Expert System] Crowe and Mutch 1994	assessment of potential for pesticide contamination	GIS, Ch DB	Pesticide transport & degradation models	integrated rule-based ES (interfacing assistance and consistency checking)
DEMOTOX Ludvigsen 1996	assessment of potential for organic chemicals conamination.	Ch DB	Organic chemicals transport & degradation model, ranking method	rule based interfacing assistance (contains heuristic knowledge)
DSS for evaluation of pump-and treat alternatives, Rifai et al 1995	evaluation of pump-and treat alternatives	Hg DB, Ch DB,	GW flow and cont. transport models	hypercard, slides, video clips, (general and site-specific knowledge)

[a] [] Terms, as used by the developers.
Abbreviations used in the table: Ch-chemical, DB-database, DSS-decision support system, EX-expert system, GIS-geographical information system, GW-groundwater, Hg-hydrogeologic, VA - vulnerability assessment.

of computer software and hardware. Appearance of Geographical Information Systems (GISs) has made the integration (especially with groundwater models) very attractive. Beside coupling of data bases (e.g. GIS) and models, work has also been done on integration of qualitative information into DSSs. Therefore, DSSs for groundwater quality management are nowadays usually seen and described as software tools that integrate data bases, groundwater models and knowledge bases (Table 2.1). Firstly, several data bases have been reviewed, followed by the review of software that integrate data bases with groundwater models and vulnerability assessment methods. Subsequently, knowledge-containing software are introduced, firstly ones related directly to the groundwater modelling and then those associated with more specific GWQM tasks.

Development of hydrogeological data bases was, until recently, mainly directed towards effective data storage and browsing, while no special attention was paid to data processing and presentation. However, data processing and especially presentation are becoming more important features of the data bases. GWW (UNESCO, 1995) is a relational data base aimed at assisting in most general or standard hydrogeological tasks that involve data processing and presentation (mapping, statistical analysis, chemical analysis, etc.) A HydroGeological Data Base (HGDB) with a rather specific structure and purpose has been developed in the USA (Newell et al, 1990b). HGDB contains hydrogeological information from 400 field site investigation used to classify aquifers as one of the DRASTIC[2] hydrogeological settings. Detailed statistical summaries of five groundwater parameters are included that can be used to check the reliability of new field values, or to estimate a parameter value when the field data are not available.

The need for accurate spatial data presentation has led to the development of Geographical Information Systems. GISs are basically geographical data bases with superb capabilities for accurate spatial analysis and presentation. TNO Institute of Applied Geoscience (TNO, 1994, Deckers, 1994) has integrated a relational data base and a GIS in order to create a Regional Geohydrological Information System (REGIS). Besides a data storage kernel, REGIS has the 'application environment' consisting of a set of geohydrological applications that cover processing and analysis of a specific type of information. The object-orientated architecture of REGIS provides considerable flexibility and adaptability of the System. REGIS is coupled (or extended) with toolboxes (e.g. Statistical or Graphical) used for additional data processing and presentation.

The coupling of Geographical Information Systems and groundwater models has become more or less common practice. Integration speeds up considerably the preparation of input data for groundwater models and graphical analysis of model results (Walsh 1993, Furst et al, 1993). GISs are usually coupled with groundwater flow models to enhance pre- and post-processing

---

[2] DRASTIC (Aller et al, 1987) is a standardised system for evaluation of groundwater vulnerability using hydrogeological settings - mappable units based on hydrogeological factors that control groundwater flow. These factors, which form the acronym DRASTIC, are incorporated into a relative ranking scheme that uses a combination of weights and ratings to produce a numerical value called the DRASTIC index .

model operations (Biesheuvel and Hemker, 1993). A GIS could be more actively involved in model calibration through a combination of visual and relational querying (Roaza et al, 1993). In the latter case the ARC/INFO GIS has been integrated with a groundwater transport model (SWICHA code). A user interface has been developed that integrates a SYSTEM 9 GIS and a groundwater model for delineating Well Head Protection Areas (WHPAs) around public water supply wells (Rifai et al, 1993).

There are a several examples of coupling of a GIS with the DRASTIC scoring system (Evans and Myers, 1990, Rudnquist et al, 1991, Padget, 1994, etc.). The integration ranges from rather loose to full. It seems that GISs and DRASTIC have already been formed into a standard package for groundwater vulnerability assessment in USA.

However, there are very few examples of the integration of any kind of knowledge component into DSSs, in spite of continuing calls for integration (e.g. Fedra, 1990). An expert system has been developed and integrated with a GIS to assist in the definition of input parameters and 'decision variables such as source strength or pumping rates' (Fedra, 1993). For example, after inquiring about the size of an irrigation area, technology, management and crops, the system will use simple models or rules to estimate (and to suggest to the user) crop- and irrigation water demands. A specification of input parameters is allowed only within a certain, plausible range. The system contains ICAD (Interactive Computer Aided Design, (Fedra and Dierch, 1989)) that alleviates work with the GIS by setting in motion adequate queries. The rules embedded into the system can be displayed during the user-system dialogue. That provides transparency of the system. The hypertext that contains explanations of the terms used in the rules also contributes to the system's transparency.

All the advantages of integrated information are demonstrated in 'OASIS', a decision-support system for groundwater contaminant modelling (Newell et al, 1990a). OASIS combines databases, groundwater models and reference libraries into an user-friendly environment that is built up using the HyperCard (Apple Macintosh) environment. HyperCard is a relatively new type of information medium wherein large amount of diverse information is organised and presented in separate screens or 'cards'. Various types of links are provided to connect cards into groups called 'stacks'. OASIS has more than 1700 cards connected in two dozen stacks. The architecture of OASIS follows a 'process oriented' approach to the modelling (Loucks et al, 1985) where use of integrated tools rather than single complex models allows active participation of non-modellers in the modelling procedure. OASIS contains two chemical databases, a hydrogeological database HGDB (here called DRASTIC) and the EXPAR knowledge base. EXPAR was developed to assist in preparation of input parameters for analytical contaminant transport modelling (McClymont and Schwartz, 1991a,b). In its original form, EXPAR was coupled with a analytical model into the ROKEY Expert System organised in several elaboration and assistance programs. The elaboration programs provided information on transport parameters, underlying processes and lists of references. The assistance programs were rule-based and assisted a user to derive parameter values. Prior to a model run, a consistency checking of parameters was carried out. ROKEY is written in Fortran and does not include any graphics.

Rapid development of model codes has created the necessity to integrate information on their availability and applicability. In 1992, the International Groundwater Modelling Center (IGWMC) built up a database of groundwater models called MARS (Model Annotated Retrieval System). The same year, KGM (A Knowledge Based System for Groundwater and Soil Pollution Modelling) was developed as a result of cooperation between several European institutes and universities (Saaltink and Carrera, 1992). KGM guides the user through the problem analysis that yields a conceptual model of a problem. Elements of a conceptual model are then compared with characteristics of the model codes. As a final product, KGM offers a list of model codes ranked according to their resemblance to the conceptual model. Unlike KGM, MODEL EXPERT95 (Scientific Software Group, 1995) is a commercial product meant to provide more practical information on model codes such as price, availability, number of users, etc.

Several expert systems have been developed to assist in assessment of a groundwater pollution situation. GEOTOX (Wilson et al, 1987) is a rule-based knowledge system for identification and ranking of waste disposal sites. The factors used in GEOTOX for the assessment of site characteristics are divided into three groups: *Permanent Hazard* includes factors which describe the hydrogeological/soil conditions (environmental setting) of the site and possible pathways of contamination; *Local Hazard* describes a contaminant, while *Global Hazard* includes possible targets or receptors of contamination. Each site characteristic is associated by a set of expert rules that define its contribution to the overall hazard. The importance of site characteristics is proportional to the weight derived by expert judgement and to the user's confidence in the reliability of data. The interference engine in GEOTOX attempts to consider al possible interaction between site characteristics. After a priority list of characteristics is defined, a score system is introduced to rank site hazard. A prototype of an expert system for waste disposal site selection has also been developed by Rouhani and Kangari (1987).

A methodology for identification of unknown sources of pollution based on the concept of statistical pattern recognition has been embedded in an expert system (Datta et al, 1989). The function of the pattern recognition method is to match statistically an observed set of concentrations in the field with a comparable set obtained by simulating groundwater transport for various disposal conditions. The expert system uses the results of matching to select particular pollution source locations and magnitudes.

The EXPRES expert system (Crowe and Mutch, 1994) is intended to be used as a screening tool by non-experts who need to evaluate the potential for pesticide contamination of ground water. It combines a knowledge-based system, a graphically-based user-system interface, a pesticide chemical data base and three pesticide assessment models. Expert knowledge, a chemical data base and a transport/degradation model are also integrated in the DEMOTOX expert system (Ludvigsen et al, 1986). The core of the DEMOTOX is a organic pollutant ranking model which utilises a mobility and degradation index (MDI). Final MDI values are obtained by multiplying calculated values by the expert's confidence factors.

This review will be rounded off with a note on DSS for groundwater remediation (Rifai et al, 1994). Three modules (Global, Site Specific and Simulator) are integrated into a DSS that assists the user in evaluating pump-and-treat remediation alternatives. The Global module comprises data bases and reference libraries, while the Site Specific module uses documents, slides and video to presents a real-world case-study. The Simulator module contains a number of mathematical models for design and analysis of pump-and-treat schemes. The DSS is designed in the same way as OASIS, and developed for both the Macintosh and PC Windows environment.

2.3 Groundwater quality management in the light of knowledge encapsulation

Users and the utilisation. Groundwater quality management comprises the whole set of technical, institutional, legal and operational activities. Accordingly, one can expect that groundwater problems are managed by teams of specialists, or by a groundwater manager who coordinates the work of various specialists. From a practical point of view, a profile of a 'groundwater manager' cannot be uniquely defined, being dependent on factors such as type of groundwater problems, country (political, economical, social issues), size (a level of managing) and a type of company (consultancy, water supply company, governmental institution), etc. Nevertheless, groundwater problems are by definition hydrogeological problems and the vast majority of the work is carried out by hydrogeologists[3]. Various software tools have been developed to assist a hydrogeologist in her/his work. As the tools (e.g. groundwater models) become more sophisticated, the problem of their poor and/or inadequate use emerges clearly. The problem extends far beyond running the software; it is very much related to input data analysis, as well as the interpretation and documentation of the results of data processing. Integration of software tools speeds up processing of quantitative (numerical) information, but complementary qualitative information (knowledge) is needed to enhance analysis, processing and interpretation. In an ideal case, the user would possess all the knowledge required, but even then knowledge integrated into a DSS would be an useful reminder.

Limitations and obstacles. The type of GWQM problem defines substantially the requirements for the analysis, processing and presentation of information, and therefore the various software tools that might be required to be integrated into DSS for GWQM. By adding the fact that each problem type requires some specific knowledge, it becomes difficult to imagine how universal DSS could be developed. However, if the problems are classified according to the source of pollution, two global types can be distinguished: local and regional. The majority of sources are of a local type (point-source pollution), while agriculture is recognised as a prime cause of

---

[3] The distinction between a groundwater manager and hydrogeologist is basically a matter of (level of) specialisation and authority of decision-making.

diffuse, regional pollution. As point-source pollution problems show many similarities (with respect to posed goals, estimated parameters, applied methods, etc.), the scale of the problem is eventually the only factor which defines global content and design for a DSS for GWQM.

The main difficulty regarding knowledge encapsulation is, however, not the scale, but the case-specific nature of groundwater pollution problems, defined by their hydrogeological background: there are no two identical groundwater systems and no two identical groundwater pollution situations. GWQM problems ask for case-specific handling that, successively, demands case-based reasoning and use of 'specific' knowledge. This issue will be discussed in detail in a later section.

Application realms. 'Groundwater pollution' is the key-word in groundwater quality management. Assessment and prediction of a groundwater pollution situation, groundwater protection from pollution and remediation of polluted groundwater, are the main groundwater quality management tasks. Selection of protection and remediation measures is based on assessment and prediction of the pollution situation. Among various methods used to assess and predict groundwater pollution, groundwater modelling and groundwater vulnerability assessment can be distinguished as being the most important. In groundwater pollution studies, site characterisation is (or should be) considered as an independent task that precedes all the other, above stated, tasks. Issues related to knowledge that is required for site characterisation, groundwater modelling, vulnerability assessment and remediation of groundwater pollution, will be discussed further below.

2.3.1 Site characterisation

Site characterisation includes basically data collection, processing and presentation. Those tasks are carried out as field work, laboratory analysis and as desk-top studies. The latter comprises preparation of field and laboratory activities (planning a campaign) and analysis, processing, presentation and interpretation of the results obtained. A basic taxonomy of site characterisation that can be carried out with the assistance of a DSS is given in Figure 4.3.

Analysis. Tasks to be carried out in order to solve groundwater pollution problems are described in numerous protocols, guides, manuals and similar documents, usually dedicated to a certain type of problem. These documents are very often too general, extensive and muddled and therefore rarely used. A protocol has to resemble a problem structure in a clear way, starting with the most general items related to pollution problems (e.g. site characterisation, pollution prediction, or remediation) and then proceed to more detailed level for each particular item (e.g site characterisation→ pollution source characterisation → source delineation→ lateral delineation → methods). Relations among objectives of a study, parameters to be estimated and methods to be used have to be displayed in a transparent manner, regardless of their complexity.[4] Despite their specifics, groundwater pollution problems show much in common, regarding posed objectives, performed tasks, estimated parameters and applied methods. In order to encapsulate this generic knowledge into DSS, ordering of GWQM tasks and associated knowledge is required. Electronic encapsulation can provide efficient on-line browsing of protocols (using hypertext hotspots, indices, glossaries, etc.) and direct links between various issues (objectives, parameters, methods) needed for the analysis.

The purpose and the content of site characterisation are defined by the objectives of characterisation. Development of conceptual models of the groundwater system (i.e. system matrix, boundary and internal conditions) and the characterisation of the groundwater pollution situation (delineation and chemical characterisation of contaminant source and contaminant plume) are the general objectives of site characterisation. Information on these objectives can be worked out according to the type of pollution problem and electronically linked with information on corresponding parameters and methods.

Two kinds of information on parameters can be embedded into the DSS: (1) basic information, to serve as a reminder, that would also include default values or ranges; (2) specific information that is obtained during the previous assessments, supported by description (of factors involved) and the possibility for augmentation.[5] Finally, procedures for checking consistency and the range of the parameters can be built into the DSS.

The methods are used to obtain data that are stored in databases. The methods are also used within a DSS (by the integrated software) to process the data. Information on the first group of methods (mostly field, and some lab methods) could be used as a permanent reminder on method characteristics, target and design parameters, measured quantities, geological, hydrological and other constraints, scale of application, uncertainty, easy of applicability, costs, etc. (Peck et al, 1988). Experiences in use of the methods should be included as well, together with the possibility left to the user to add his/her new experiences.

---

[4] One parameter can be required for fulfilling more than one objective; some parameters can be determined by application of various methods; likewise, some methods can determine (or indirectly contribute to the determination of) more than one parameter, etc (Figure 4.3).

[5] Not all the parameters are experimentally defined for each groundwater pollution problem, so the parameter value has to be taken from 'literature'. Usually there is no clear default value, so information on all factors involved is needed to secure correctness of selection. When a few dozen values are available (e.g. for dispersivity) with extended adjacent information on factors involved (aquifer material, average aquifer thickness, hydraulic conductivity, effective porosity, velocity, flow configuration, monitoring, tracer, method of data interpretation, scale of test and classification of reliability (Gelhar et al, 1992)), then encapsulation of the knowledge is more than desirable. New tests are continuously carried out, so an option that provides augmentation of the encapsulated knowledge is necessary.

Processing. DSS modules (e.g. a GIS or statistical toolbox) contain a number of methods for spatial and time series analysis. Brief, experience-based and carefully selected information on methods used to process numerical information within the DSS would promote and enhance their use. Information to be processed defines the kind of processing to be performed. Various procedures (queries and special applications) could be accordingly developed in a DSS to automate the processing.

Presentation As a result of application of the methods, numerous maps, profiles, diagrams, charts and tables are created. According to the posed objectives, estimated parameters and applied methods, various forms of presentation could be pre-designed and encoded in a DSS to provide the most appropriate presentation of the acquired (and processed) information. The forms should be, as much as possible, in compliance with (input requirements for) methods used for the assessment (e.g. models).

Interpretation. Availability of integrated background information required for pollution problem analysis and availability of integrated procedures for consistent and adequate processing and presentation, creates an environment for correct site characterisation. Complete (qualitative and quantitative) information on site characterisation has to be stored in the DSS in such a way as to encourage its reuse. For that purpose the DSS could contain a 'reporter' module, a kind of text processor with a protocol that would assure (or at least enhance) the quality of the reporting and documenting.

2.3.2 Groundwater modelling

Most serious errors in modelling are related to the first few steps of the modelling protocol (Anderson and Woessner, 1992), namely: selection of an appropriate model, model conceptualisation and estimation of input parameters.

Selection of the appropriate model comprises the selection of: modelling technique, model code and model complexity. The taxonomy of the groundwater models surveyed more than decade ago (van der Heijde et al, 1985) counted almost 400 modelcodes designed for identification of groundwater systems, prediction of groundwater flow and solute transport, and management practice. Selection of a modelcode receives nowadays appropriate attention; besides extensive overviews of model codes (Quercia, 1993, Van der Heijde and Elnawawy, 1993), databases (IGWMC, 1992), and even expert systems (KGM and MODEL EXPERT) are available to assist in the selection procedure. In KGM the selection is based on the conceptualisation process and that makes this scientific software very attractive. It would be worthwhile to complete the development of KGM by enlarging the number of model codes and by including information on code verification[6]. The expert systems for selection of the model code can be used independently, hence without the integration into a DSS.

Modelling techniques are widely covered in the literature (e.g. Kinzelbach, 1986, Bear and Verruijt,1987). Since the modelling techniques form the basis of the model codes, expert systems for selection of model code might also include an overview of the techniques followed by their comparison. DSSs for GWQM should contain information on integrated model code(s) as well as the modelling technique applied in the model code(s); that would include the processes approximated by the technique (e.g. dispersion), consequences of the approximation (e.g numerical dispersion) and the model code parameters (e.g. Peclet criterion). The experiences in using the model code could be added as well (e.g. Zomorodi, 1990). A substantial part of information on processes and parameters can be (within a DSS) 'inherited' from the site characterisation.

The selection of model complexity is, in principle, based on the objectives of the modelling study, the complexity of the groundwater system and data availability. However, an undue sophistication or so-called 'overkilling' has often been noticed in practice (e.g. Reddi, 1990), where model complexity does not portray the factors it should be based upon. (On the contrary, application of simple analytical models always proves to be useful, both in preliminary groundwater studies, and as a control of complex numerical models.) There is little that can be done to reduce overkilling at the current state-of-art of groundwater modelling. Adequate processing of information (integrated into the DSS) during the site characterisation, would assist the user in making a better judgement on (and trade-off between) objectives, problem complexity and data availability, these being the main factors that define model complexity. Besides, the examples of 'appropriate' model use could be made available to the user in a form of modelling case-studies or peer reviews.[7]

A translation of hydrogeological information into *a conceptual model* suitable for numerical modelling is the most difficult and very often the most underestimated step of the modelling protocol. There are very few firm rules (and no recipes) that can be applied in order to avoid misconceptualisation. Usually more efficient and more careful processing of collected data (desk-top analysis) is recommended (McLaughlin and Jonson, 1987, Grondin et al, 1990), showing once more the importance of site characterisation. KGM is the only knowledge-based system identified to date which exercises a conceptual approach. However, the conceptualisation in

---

[6] Code verification is already for years 'a hot issue' within the research community. The situations that numerical models are built to deal with (e.g. heterogeneous system properties and irregular boundaries) cannot be validated. However, some validation (e.g. using the so-called 'walk-through') is possible and the results should be made available to the user of the ES.

[7] In the FEFLOW system (Fedra and Diersch, 1989) two kinds of modelling cases are offered to the user, namely: generic, such as bank infiltration, deep well injection or landfill; and specific, referring to the real-world cases already carried out. The cases can be modified according to the specifics of a new problem. However, only the quantitative information on cases is available.

KGM is carried out in very broad terms, covering all kinds of modelling problems (unsaturated and saturated zone, flow and transport, density dependent flow, etc.) that occur in all kinds of hydrogeological environments. Any attempt to make a conceptualisation more specific and useful not only for the selection of a model code, but also for the development of a numerical model, would be hard and most likely futile work[8].

The view on the content of conceptualisation is not unique; for further discussion see e.g. Bear and Verruijt, 1987, National Research Council, 1990, Anderson and Woessner, 1992. Nevertheless, the parameters that represent properties of the groundwater system as well as processes that govern groundwater flow and transport have to be estimated[9]. One way to structure information on the model parameters and to embed it into the software is demonstrated in ROKEY, the expert system described in the previous section. The breath-taking development of computer software engineering has made ROKEY (within ten years) an ancient product. However, for the development of a knowledge-based system in hydrogeology, ROKEY is a valuable prototype that needs an appropriate graphical user interface (GUI) to become a modern knowledge-based modelling tool. The knowledge embedded in ROKEY supports input preparation for an analytical model. The same information (with some modifications) can also be used for numerical modelling; the question is: where can the knowledge needed for complete set up of a numerical groundwater model (i.e. including geometry, boundary and state conditions, network density, etc) be found? There is very little advice of general type that could be offered to the user, such as: (1) the boundaries of the model should be located along natural limits if possible, or well away from region affected by pumpage or recharge, or (2) the influence of boundary conditions should be checked by changing head to prescribed flux and vice versa.

Apparently, there is (almost) no knowledge on groundwater modelling that can be formalised in rules. One of the reason for that is the case-specific nature of groundwater modelling problems. Nevertheless, there are cases which are similar, so the solution for a 'new' problem could be found by browsing through the 'old' similar cases and adjusting the old solutions[10]. There are a few obstacles to efficient use of modelling case-studies within an DSS. Guidance for documenting the modelling process is badly needed, as well as quality assurance (QA).

---

[8] The full name of the KGM is a 'Knowledge Based System for Modelling Soil and Groundwater Pollution' (Saaltink and Carrera, 1992). It is, according to the background documents, aimed at giving guidance, information and advice in the different stages of the modelling process. In practice, however, the KGM cannot offer more than assistance in the selection of model codes. The main achievement of KGM is in structuring and encapsulating knowledge on groundwater modelling.

[9] *Parameter estimation* is seen here as the estimation of model input, based on field data and on results of the site characterisation. Parameter estimation is *not* used here as being synonymous with model calibration (which is again synonymous with solving the inverse problem) which is often the case.

[10] ROKEY provides the user with the option to use a parameter value obtained from 'similar site'. This concept of 'similarity and comparison' could perhaps be extended to the other, more complex issues (e.g. type of problem, management practice, hydrogeologic conditions), or on combinations of those issues, presenting modelling a case study as an quantitative (like FEFLOW) and qualitative entity.

Documentation is important to ensure that the study could be reliably repeated, while QA is crucial to both development and application of the model (National Research Council, 1990). As long as the modelling case studies do not fulfil these requirements, the systematisation and encapsulation of 'case based' knowledge will be extremely difficult. The use of case-based knowledge has a strong endorsement in a cognition theory called 'case-based reasoning' (CBR). CBR will be addressed in more detail in Section 2.4.

Modelling has become the main method used in groundwater studies, and certainly not without reason; models simulations and predictions are considered to be a very important source of information for decision-making in GWQM. Therefore an 'appropriate' use (appropriate with respect to issues addressed in this section, and others) of groundwater models is always welcomed, as long as it yields 'reliable results'. (It seems that model results always look more reliable than they really are ?!) A lot of effort has been made in order to improve the reliability of groundwater models, especially through (automatic) calibration (Yeh, 1986, Carrera and Neuman 1986, Olsthoorn, 1995, 1996, etc), where some new optimisation approaches (e.g. genetic algorithm) has recently been introduced (see Section 2.4). However, the main issue is validation of the model results. Post-audit seems to be only reliable validation procedure (e.g. Deak et al, 1996), although not all modellers agree. Instead of entering into discussion on this issue here, the famous (and provocative) statement of Konikow and Bredehoeft will be quoted: 'Groundwater models cannot be validated', to encourage the reading of their article (Konikow and Bredehoeft, 1992) and the following discussion.

2.3.3 Groundwater vulnerability assessment

The concept that some areas are more likely than others to become contaminated has led to the use of the terminology 'groundwater vulnerability' to contamination. The concept is based on assumption that the physical environment may provide some degree of protection to groundwater against natural and human impacts, especially with regard to contaminants entering the subsurface environment. Vulnerability assessment (VA), however, goes beyond the basic concept of groundwater vulnerability; it often involves potential targets of groundwater pollution and elements of risk analysis. It would be quite difficult to give a precise definition of vulnerability assessment. Nevertheless, according to the recent reviews (National Research Council, 1993; Vrba and Zaporozec, 1994), it covers much more than the sole application of vulnerability maps.[11] Here, a brief overview of various (often knowledge-based) vulnerability methods will be given with respect to their possible integration into the DSS for GWQM.

---

[11] Vulnerability mapping is the approach of vulnerability assessment that has experienced a rapid development within the last decade (Van Duijvenboden and Van Waegeningh, 1987), mostly due to the development of geographical information systems.

The most general classification of VA approaches (National Research Council, 1993) includes: (1) methods employing process-based simulation models, (2) statistical methods, and (3) overlay and index methods.

(1) No particular reason could be found to include simulation models in VA approaches. The Council considered only the unsaturated flow and transport models that are used to estimate a contaminant potential of diffuse pollution sources (agriculture). Integration of these models in a DSS is, however, highly recommended. Since models like CMLS, GLEAMS, LEACHM, etc. require extensive use of chemical data, an effective coupling with a chemical database is very beneficial to the user (e.g STF chemical database is coupled with the VIP and RITZ models (Sims, et al., 1991)). Further integration with a GIS and one of the vulnerability mapping methods (e.g. DRASTIC, see the software review) would lead to development of a DSS for regional groundwater pollution problems.

(2) Statistical techniques for vulnerability assessment reviewed by NRC are those of general type, such as: regression, analysis of variance, discriminant and cluster analyses, geostatistics and time series analysis. Some application of classical statistics in regional pollution problems can be found in literature (e.g. Teso et al, 1988). However, it is not yet apparent what special benefit could a vulnerability assessment (especially at a local scale) gain from the use of various statistical methods. In that respect, the expert system 'Embedding Pattern Recognition Technique' (see the software review) is a promising undertaking.

(3) Assessing methods in the third group, overlay and index methods, are based on combining maps of various parameters (e.g. soil, hydraulic conductivity, depth to water table) of the region by assigning a numerical index or score to each attribute. This group of methods shows a considerable resemblance with the Parametric System (PS) methods as classified by Vrba and Zaporozec (1994). The PS methods are further divided into: matrix systems, rating systems (e.g. LeGrand's system) and point count system models (e.g. DRASTIC). These methods were developed for vulnerability assessment to be carried out on a local or regional scale. LeGrand's system is, for example, aimed at assisting in evaluation of waste-disposal sites, being a typical local scale problem (LeGrand, 1980). Most of the VA methods are, however, meant for regional assessment (e.g DRASTIC). Their suitability for assessment on a local scale is conditioned by parameter variability which decreases with scale, so that differences in the parameter value become quite irrelevant at a local scale.

As the majority of the 33 types of groundwater pollution sources (as identified by US Office of Technology Assessment) are of local type, the bulk of modelling work (especially that of the saturated zone) and groundwater remediation are related to point-source pollution problems. Apparently, a vulnerability assessment method is needed which would (based on gained experience) incorporate best of the existing methods (LeGrand system, GEOTOX, Hazard

Ranking System, etc.) and fully utilise the advantages of integrated information (e.g. results of site characterisation) and integrated software (GIS, databases, Statistical toolbox)[12].

2.3.4 Groundwater remediation

In accordance with increased interest in the protection and remediation of groundwater, several publications have been produced to cover this topic (e.g. Rail, 1989, Charbeneau et al, 1992, Bedient et al, 1994). Yet, not enough has been done in the direction of completion of a comprehensive and systematic overview of protection-remediation methods. That one of the best overviews found in literature has been produced by an consultancy (IWACO, 1995), shows the extend of the practical demand. Hydrogeologists need the overview in order to make a first step in groundwater remediation, i.e. select a method. This need is correctly recognised by developers of the DSS for evaluating pump-and treat remediation alternatives (Rifai et al, 1994).[13] This rather comprehensive DSS contains a library with short summaries about various treatment processes commonly applied in soil and groundwater remediation. Curiously enough, the DSS includes a complete, vividly-presented example of a real case study. The case-specific nature of groundwater problems and the state-of-the-art of knowledge on remediation processes make extraction of 'relevant' experience (in a form of rules, for example), that is gained by conducting case studies, a very difficult task. IWACO (1995) compiled twenty four case-studies on GWQM carried out throughout Western Europe. Such compilations are not useless, but the best way of transferring the experience to the user is to present the information on the pollution problem as a whole (the case study). Contemporary hardware can accommodate large amount of information (e.g. a pile of case studies) which, however, has to be encapsulated in a systematic and uniform way to allow its efficient browsing and reuse (see case-based reasoners in next section).

2.4 Artificial Intelligence (what does it offer ?)

Artificial intelligence (AI) is (the automation of) activities that we associate with human thinking, activities such as decision-making, problem solving, learning...(Bellman, 1978). This is just one of numerous (and not necessary mutually-agreeable) definitions of AI. Nevertheless, in its most genuine form AI deals with human thinking and reasoning, and with knowledge which emerges therefrom. Hence, before attempting to encapsulate knowledge artificially, one ought to become acquainted with the basic postulates of AI related to human cognition and corresponding knowledge representation.

---

[12] Could the overlay approach, at least for some parameters, be adapted for the local scale assessment?

[13] DSS for evaluating pump-and-treat remediation alternatives is seen as an important contribution to development of DSS for GWQM, especially concerning the information it contains. Integration of a GIS and work on virtual integration and interaction of (qualitative and quantitative) information within the system could be considered for further development of the system That would alleviate design of a remediation system and enhance evaluation of remediation schemes.

2.4.1 Reasoning and knowledge representation

Reasoning. According to the traditional view, human thought depends upon a set of reasoning principles that are independent of any given domain knowledge. Artificial Intelligence therefore attempts to represent knowledge as generally as possible by formulating rules which are widely applicable (concept of the general problem-solver). If solving a problem means following a set of general rules about problem-solving, then any domain of knowledge can be captured by a set of rules about problem-solving in that domain. Such rules represent domain knowledge and attempts are made to encode the domain knowledge in rules that will be used by an expert system.

All the methods and techniques for *knowledge representation* developed within AI, with the exception of a case-based reasoner, are rule-based. Even the most recent approach that introduces a concept of intelligent agents (Russell and Norvig, 1995), does not depart from the traditional rule formalism. The rules are derived using the principles of logic, i.e. syntax, semantics and deduction[14]. So-called *production rules* (simply 'If-Then statements) are the most widely-used structures for knowledge representation. The inferences are made by using forward or backward chaining. Rules are applied in forward chaining to derive new facts from those that are already known. In backward chaining, a conclusion is made in the first place; then recursive inferencing is applied to define whether conditions are met that allow such conclusion. If set of rules condition a situation (or an action to be taken), it is convenient to combine those rules in so-called decision tables (Table 2.2). The rules and their organisation are comprehensively described in a number of books (e.g. David and King, 1977, Weiss and Kulikowski, 1984).

Table 2.2 Decision table, an example

STATE CONDITIONS			
Time regime	Transient		Steady state
Initial conditions	known	not known	-
State conditions OK	YES	NO	YES

Knowledge in some domains is fairly structured. If not, a structure has to be imposed, to allow processing and representation of a large amount of knowledge. Structuring of knowledge is done by the use of semantic networks, frames and object-orientated models. The frames are, in fact, extension of semantic networks, and object-orientated models are extension of frames. The concept of *semantic networks* is based on the premise that the human memory functions in part

---

[14] The syntax of a language describes the possible configurations that can constitute a sentence. The semantics determine the facts in the world to which the sentence refer. Deduction or logical inference is a process that implements the entailment relation between sentences. The entailment relation is established if a new generated sentence is true, given that the old sentences are true. The entailment mirrors the real world relation of one fact following from another.

with associations between objects, concepts and events (Goodall 1985, Waterman, 1986). A segment of a hypothetical network is illustrated in Figure 2.1. The two main components of the network are nodes (e.g. chloromethane) and arcs (e.g. 'is-a') that link the nodes. Other commonly used arcs are 'has', 'owns', 'needs' and 'is-part-of'. The main characteristic of semantic networks is inheritance; chloromethane is, for example, both a 'HAH' and a 'Toxic Chemical'. An inheritance characteristic is used in rule-based systems to reduce the number of rules needed in the knowledge base[15]. The nodes in a semantic network can be represented by *frames*.

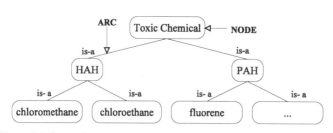

Figure 2.1 Semantic network, an example

A frame contains a group of related parameters called slots (Figure 2.2). Each slot has an attribute name that gives a characteristic of the concept represented by the frame. In addition, the slot contains default values, plausible values, and optionally possible error values and/or measures of importance. The frames provide a means of grouping similar objects, events and attributes, being very useful for representing the knowledge in domains that are characterised by well-established taxonomies. Structured knowledge representation by frames was originally introduced by Minsky (1975).

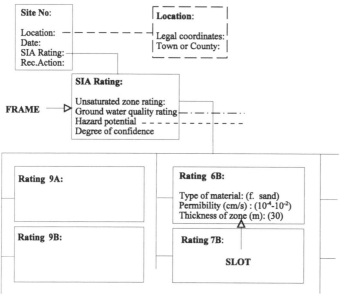

Figure 2.2 Frame-based system, adapted from McClymont and Schwartz, 1987

The fundamental difference between frames and *objects* is that the latter, besides attributes, contain also procedures (or operations), meant to describe the behaviour of the objects. Until recently, a term 'object-orientation' was related exclusively to the so-called object-orientated programming languages, such as Smalltalk and C++. Rumbaugh (Rumbaugh et al, 1991) brought object-orientation from

---

[15] For example, IF: the compound is-a Toxic Chemical and the facility overlies a shallow aquifer and leakage has been detected THEN: the disposal facility is a hazardous waste site.

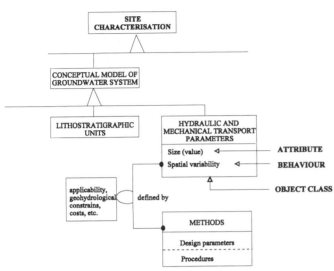

Figure 2.3 Object-orientated structure, an example

programming up to the modelling level. The object-orientated approach is based on modelling objects from the real world and then using the model to build a design organised around those objects. The fundamental construct is the object, which combines both characteristics (attributes) and behaviour (operations) of the objects (Figure 2.3). Objects with the same attributes and operations are grouped into a class. In terms of GWQM, wells, aquifer systems, river channels, etc. can be considered as objects or object classes.

REGIS (see software review) is only piece of hydrogeological software identified to date that has been developed using the object-orientated approach.

2.4.2   Case-based reasoning

Case-based reasoning. First ideas about case-based reasoning originate from Schank and Abelson (1977) which claim that the human mind is not a system of rules, but a library of experiences. They rejected the hypothesis that human thought depends upon a set of reasoning principles or rules and introduced so called "scripts", much larger amounts of knowledge, as a primary knowledge structure. Their hypothesis states that people remember an event in terms of associated scripts (going to a restaurant, taking a bus). Thinking, therefore, means finding the right script to use (if it is available), rather than generating the new ideas. Explanation, the central process in everyday understanding, is a process of adaptation rather than creation; it is possible to explain something new by adapting a standard explanation from another situation. Facing a new problem, people strive to find the best plan they have heard of or previously used that is closest to the new problem, and attempt to adapt that plan to the new situation. When a problem situation is solved, a new experience is stored in a memory, to be used again. If the same situation is repeated, gained experience has no uniqueness any more and it merges into generalisation or a new script. Created 'general' knowledge is referred as prototypical cases (Porter et al, 1990) or ossified cases (Riesbeck and Schank, 1989). They are, in principle, a set of norms or rules, but formed by aggregation of identical cases (having the large amounts of already composed knowledge as a starting point for reasoning), rather than by composition of abstract operators, based on independent reasoning principles.

General and Specific knowledge. Some pieces of knowledge representing an experience do not merge in general knowledge. They stay in a memory as separate instances, next to the general knowledge. General knowledge provides a framework for reasoning, but does not provide universal applicability of the same general piece of knowledge to any particular situation. Cases contain specific knowledge applied in a specific situation. This knowledge, that might be too hard to capture in a general model, allows reasoning from specifics[16]. Both kinds of knowledge, general and specific, are needed in problem solving.

CBR cycle. Knowledge, contained in memory in the form of cases, is used by the reasoner to explain a new situation or to create a solution to a new problem. In order to do that, a Case-Base Reasoner has to perform the following tasks (see Figure 2.4): identify the current problem situation, find a past case similar to the new one, use that case to suggest a solution to the current problem, evaluate the proposed solution, and update the memory by learning from the experience (Aamodt and Plaza, 1994). Characteristics of a new case are defined by a number of selected features or indices (Figure 2.5). Indices are used to retrieve similar cases from the memory whereas the match rules decide which case is more like the new one. The old solution is then modified and tested. If the solution passes the test, indices are assigned and the new case is stored to be reused when needed. If the solution fails, a failure should be explained, and the working solution tested again (Slade, 1991).

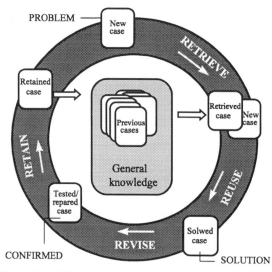

Figure 2.4 CBR cycle

Indexing problem and memory organisation. In CBR a new situation is explained in terms of the recalled experience. A reasoner has to find in memory a case containing experience closest to the new situation. It can be done if indices are assigned to cases stored in memory so that they can be recalled when appropriate. Recalling cases appropriately is the central issue of case-based reasoning and is called the 'indexing problem' (Kolodner, 1993). Indexing of cases is related to case content and organisation of case memory. Most of the CBR developed so far use the Dynamical Memory Model (Schank, 1982) to organise cases in memory. The basic unit in dynamical memory are MOP (Memory Organisation Packets), that are repositories of generalised knowledge (norms) with cases as specialisations of those general models. The basic idea is to organise specific cases which share similar properties under a more general structure. Specific and

---

[16] The terms 'specific' and 'general' are sometimes used as synonyms for respectively 'strong' knowledge (knowledge specific to a problem) and 'weak' knowledge (domain-independent generally-applicable knowledge), which is here not completely the case. General knowledge is related to a part of the problem (phenomenon) that is well enough understood to be represented by the facts (rules) and be used while dealing with the same type of problems (i.e. the same knowledge domain). Specific knowledge is primarily associated with the specifics of a problem that cannot be easily (or at all) formalised by a set of rules and afterwards used elsewhere. Therefore, in our view of CBR, cases contain both general and specific knowledge.

generalised knowledge are connected through a complex indexing web. MOP-based memory techniques view cases as objects and involve standard AI notions such as frames, abstraction, inheritance, etc.

Building a Case-Based Reasoner. There are different kinds of case-based reasoners ranging from retrieval-only systems to fully automated systems. Automated systems are able to solve problems completely by themselves, while retrieval systems interact with a person to solve a problem. The kind of a case-based reasoner is largely defined by its purpose, which is again related to a level of understanding of domain knowledge. Basically, cases serve two purposes: (1) to provide suggestions of solutions to problems; or (2) to provide a context for understanding and assessing a situation. Accordingly, a CBR could be developed to solve problems, to give suggestions, or to act merely as a database that can retrieve partially-matching cases. A rapidly-growing application of the CBR type is 'help desk systems' where case-based indexing and retrieval methods are used to retrieve cases, offering to the user large amounts of filtered information. Such knowledge bases can also be a first step towards complete application of the CBR cycle.

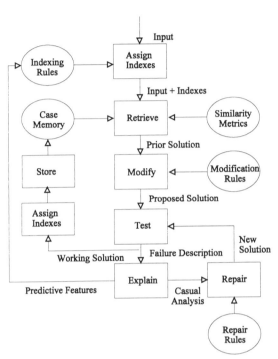

Figure 2.5 CBR flowchart, adapted from Slade, 1991

Cases versus Rules. The CBR cycle indicates some characteristics of case-based reasoning and its relation with rule-based reasoning. The biggest advantage of CBR is that solutions to problems can be proposed quickly, because they do not have to be derived from scratch. With the collection of cases stored in the memory, it is easier to adopt one which is closely similar to a new problem than to build up new solution from the beginning. CBR will retrieve cases which only partially match a new problem. The robustness of CBR enables assumptions and predictions to be made, based on experience, even if knowledge is incomplete (e.g. in domains that are not completely understood). Obviously no rules can be posed in such situations; if the problem do not match exactly any of rules, a rule-base reasoner will simply not retrieve them. Advantages of CBR in relation to rule-based reasoning can be noticed already in the process of knowledge acquisition which is often a bottleneck in constructing rule-based expert systems. The human expert could not simply make a list of all the rules to solve problem even if he is willing to do that. However, if communication is carried out through concrete examples, it is relatively straightforward and experts are usually more than willing to contribute. The next advantage of CBR is ability for learning from experience. Ability for learning is crucial for long-range success of the expert system concept in AI. CBR can augment its memory with new experiences and use them in future problem-solving. Unlike CBRs, rule-based reasoners do not have memory so they cannot remember problems that have already been solved. CBR could store in memory not only successful stories, but also failures. If mistakes are recorded, CBR could help in avoiding them.

Differences between case-based reasoning and rule-based reasoning do not mean that they 'by rule' exclude each other. There are examples of their joint use (e.g. Lopes and Plaza, 1993) and, after all, the CBR framework contains the rules. Both cases and rules are needed to handle a semi-structured problem, where information on the problem is limited and knowledge on phenomena is not complete.

2.4.3   Application of AI in Groundwater Quality Management

Numerous techniques have been developed in the field of AI to assist in knowledge acquisition, systematisation, formalisation and encapsulation. However, the genesis of knowledge-based systems for GWQM has occurred rather spontaneously to date, merely as reflection (of needs and) development of computers, and with just a glimpse of use of advanced AI techniques.[17] All the knowledge-containing software tools developed so far for GWQM have been rule-based (with an exception of OASIS, where the knowledge is organised in the form of a well-indexed electronic book). The rules in GEOTOX and DEMOTOX are structured in extensive semantic networks, while ROKEY uses a kind of frame to organise encapsulated production rules. KGM consists of a set of rule-based decision tables that is developed by using an expert system shell. The knowledge on a problem can be extracted and represented by the set of facts (rules) if the information in a domain is available and understood. Then the rule-based system can be develop presuming that a problem-solving procedure can be described by a set of well-defined steps. The main difficulty in development of knowledge-based systems for GWQM is a general lack of rules that are crucial for traditional knowledge formalisation and codification. Lack of rules is primarily a consequence of the case-specific nature of GWQM problems and the lack of formalised expertise in GWQM. An additional problem for posing the rules is that information on the problems is never complete. Lack of formalised expertise (heuristic knowledge) can be noticed in deep expert domain problems[18], as well as in managerial problems[19]. It is not claimed here that expertise does not exist. It is however obvious that it cannot be found easily, especially not in a 'ready-to-be-used' form.

---

[17] This does not hold for the information that is numerical or that can be quantified. There are numerous examples where AI techniques have been applied in groundwater management to deal with quantitative information. Specially 'self-learning' techniques, such as genetic algorithms and neural networks are attractive for optimisational GWQM problems (McKinney and Lin, 1994, Ritzel and Ehearthl, 1994, Groen and Zaadnoordijk, 1994). The topic of this article is knowledge, hence qualitative information.

[18] For example, the expertise on validity of the equilibrium sorption isotherm (with respect to flow rate) or on trade-off between model network density and number of particles (with respect to model efficiency).

[19] The problem becomes evident especially in situations where other factors then hydrogeological (e.g. ecological, social, economical) are also involved in the decision-making process.

Serious work on the taxonomy of GWQM tasks and restructuring and re-systematisation of knowledge on groundwater pollution problems is needed. The ordering of tasks according to the objectives of groundwater pollution studies, and establishing transparent connection with parameters (to be estimated) and methods (to be applied) are the prerequisite for any significant knowledge encapsulation. In his 'struggle for the soul of hydrology' Abbott (1992) states that 'hydrology is a rhetoric waiting for a grammar' (introducing loose equivalence of mathesis - rhetoric and taxonomia - grammar and genèse - dialectic)[20]. Indeed, language used in GWQM needs renewal of its grammar. A new taxonomy has to revise 'where the game is played' and 'how the rules are set'. Leaving the semiotics and semantics aside for a while, it can be stated that extensive use of the (mathematical - mathesis) methods (software tools) for processing of qualitative information, has overshadowed the importance of knowledge that enters the tools and outcoming knowledge.

Protocols, guidelines and manuals in their present form do not fulfil this expected function. US EPA has recently started to issue so-called fact sheets, documents of up to several pages that contain basic information about various tasks or processes. Such overviews are very welcomed. Owing to the inadequacy of the protocols and the attractiveness of the software tools, some important aspects of groundwater pollution problems are often ignored, and some tasks remain unfulfilled.

Neither all the knowledge needed for GWQM could, nor should, be formalised in rules. However, systematisation of acquired knowledge is badly needed. Could it be finally made clear what the generic knowledge on certain parameters, methods and processes is, what do we know for certain, where does probability help, and what comes with a big question mark? Documents like 'Subsurface Characterisation and Monitoring Techniques, a Desk reference Guide' (Eastern Research Group, 1993), work strongly in that direction. Work on taxonomy would provide a more confident approach to a new pollution problems, more authority in contesting its specifics, and in deciding what are the specifics worthwhile recording and adding next to already- acquired knowledge. If the solving of similar pollution problems (especially those of same type) demands (more or less) the same steps to be performed (protocol), the same parameters to be determined and the same methods to be applied, then generic qualitative information on these items has to be available in the most appropriate form (systematised, transparent, encapsulated). Specifics of already-solved problems (performed cases) can also contribute to solution of a new problem through processes of matching and adjustment (see postulates of case-based reasoning).

It has been already mentioned that ROKEY provides a user the option to use a parameter value obtained at a 'similar site'. FEFLOW offers quantitative information on generic and specific case studies. The DSS for evaluating pump-and-treat remediation alternatives contain a complete case study. Eventually, libraries of cases (covering the same problem type) could be encapsulated in a DSS, to support the solving of a new case. Therefore, handling of outcoming knowledge within a DSS (interpretation of results of

---

[20] Following a line of M. Foucault, Abbott (1992) states that 'every grammar presupposes a rhetoric, which is that which fills out the space of the universe of discourse with (iconical, verbal, including mathematical) signs' ('A model is in the most general sense a collection of signs that serve as a sign' (Abbott, 1993)), and continues: 'The rhetoric has again to fill out the spaces of mathesis according to taxonomia, as well as the coming to presence of these in their orderliness, their genèse.'

application of software tools, reporting and documenting) has to provide a certain quality assurance (QA) and most efficient reuse of knowledge (applying case-based reasoning). There are no illusions that an automated CBR system for GWQM could ever be developed. However, postulates of CBR as well as techniques that have been developed within the CBR research community to acquire, systematise and encode knowledge could be very helpful for encapsulation of knowledge on groundwater pollution problems.

More should be done, in general, in bridging the gap between AI and 'practice'. Development of a DSS for GWQM is, as it already stated for knowledge encapsulation, still a rather spontaneous process, without serious consideration of DSS theory. At the same time the DSS theory experiences booming development, attempting to identify scope, content and potential of DSSs. For instance, broad DSS constructors are identified (environment, task, implementation strategy, DSS capability, DSS configuration, user, user behaviour, and performance) and relations between them are subjected to investigation (Eierman et al, 1995). It is recognised that the support given by DSSs is, up to now, mainly concentrated on 'a low cognitive level' (data storage and manipulation, hypertext, consistency checking, statistical analysis, visualisation). However, it is expected of DSSs to address higher levels on application problems, such as finding (hidden) dependencies in data or helping users to find their preferences in a multi-attributive decision situation (Radermacher, 1994). The importance of learning from experience and adaptation (see CBR postulates) is often stressed, whereas the simulation of evolutionary processes is still seen as the right way of development of really 'smart' systems. It could be expected that intelligent, knowledge-based agents (e.g. Russel and Norvig, 1995) will carry out the task within DSSs and communicate with the user. Agents reason logically, taking into account their own characteristics and the properties of the (most likely object-orientated) Environment. Agents interact (cooperate), and that has led to a recent development of so-called Distributed Artificial Intelligence (e.g. Avouris, 1995).

Any more efficient and more significant application of AI in the development of DSS for GWQM (including knowledge encapsulation) requires: (1) work on the taxonomy of GWQM tasks, systematisation of acquired knowledge, and acquisition and formalisation of heuristic ('expert') knowledge, and (2) openness towards new developments in AI and computer science in general.

2.5 Concluding comments

Due to the development of computer science much has been done in the electronic integration of software tools used in groundwater quality management.Less pretentious, but still important work has been done in electronic encapsulation of knowledge needed for GWQM. Integration of the software tools and encapsulated knowledge will lead, eventually, to the development of a Decision Support System for Groundwater Quality Management.

A DSS for GWQM is seen as a system that contains tools for adequate storage, presentation, reuse and augmentation of complete (quantitative and qualitative) information on groundwater pollution problems, It has to support not only processing of quantitative (numerical) information, but also the cognitive processes, such as problem analyses and interpretation of results. Such a DSS is currently under development, and will contain modules specifically devoted to site characterisation, vulnerability assessment, pollution modelling and groundwater remediation, as discussed in Section 2.3 above. Each of these topics will form the subject of a separate article, which will address these broader issues in more detail.

It has been recognised that a case-specific nature of groundwater pollution problems and a lack of formalised expertise are the main obstacles for more significant encapsulation of knowledge for GWQM.

Protocols, guidelines, manuals and similar documents used in GWQM do not, in their present form, fulfil their expected function. Work on the taxonomy of GWQM tasks, accompanied with systematisation and formalisation of knowledge on groundwater pollution problems is badly needed. There is a strong belief that this work will open new frontiers for development of the DSS, especially with respect to application of Artificial Intelligence.

Artificial Intelligence offers a strong theoretical basis for handling large amounts of information on semi-structured problems (Case-Based Reasoning). Encapsulated information has to be transparent and exchanged among DSS components in an automatised manner. Hopefully, the concepts of intelligent agent and of object-orientated environments will contribute in designing such DSS for Groundwater Quality Management.

3. DEVELOPMENT OF A DSS: MAIN CONSIDERATIONS AND FRAMEWORK

3.1 Introduction

The overview given in previous chapter pointed out a number of problems related to DSS development. Especially, knowledge encapsulation appears to be a complex and difficult process. It has been stated (p. 32) that 'development of a DSS for GWQM (including knowledge encapsulation) is still a rather *spontaneous* process', meaning that, at this stage, no prescription can be given 'how to build a DSS'. The overview was written during 1996 and could be regarded as outdated, bearing in mind the booming development in this field. The quoted statement is, however, still valid, and the same holds for two 'concluding remarks', or rather requests, that were made relating to:

– work on knowledge taxonomy, and
– 'bridging the gap' between AI and 'practice'.

Spontaneous development of DSSs and 'DSSs' in practice has basically been governed by the need (for a such DSS) and feasibility (of its development). Accordingly, the vast majority of DSSs developed so far integrate - DSSs are almost by definition 'integrated software tools'- software for storage, processing and/or presentation of numerical (quantitative) information. On the other hand, there are very few DSSs developed that contain any knowledge component (qualitative information). This situation is primarily a reflection of the state-of-the-art of knowledge taxonomy in GWQM (it should not be too difficult to encode knowledge that is already acquired, systematised and formalised). It cannot be said that nothing has been done in this field ('responding' to the first request stated above). On the contrary, valuable books, guidelines and similar documents have been produced. Electronic encapsulation of knowledge poses, however, additional requirements, that will be discussed below, in the section on the DSS knowledge component. Interestingly, the major advance has been made in processing (and ordering) of already electronically encapsulated information. This has happened primarily due to rapid development of AI and Information Technology in general. Accordingly, 'response' to the second request (see above) has been much stronger and more successful. This holds for knowledge encapsulation, but especially for software integration and organisation within a DSS.

DSS developers are spending nowadays much less time and effort on, for example, software compatibility, and much more on DSS content and structure. That means that DSS development should gradually become a (less spontaneous and) more defined and established procedure. The recent work done in this direction is briefly discussed in the next section. Again, emphasis is on integration, and, almost exclusively, integration of quantitative information. As a DSS should, however, integrate both quantitative and qualitative information, the concept of an integrated information system is also discussed (Section 3.3). Section 3.4 describes the framework of the DSS for Groundwater Pollution Assessment. The DSS is a result of effort made to apply the concept of integrated information, as given in Section 3.3. Accordingly, most of the work done has been related to knowledge encapsulation and development of the DSS knowledge component. Section 3.5 contains general considerations on knowledge encapsulation, knowledge processing and the DSS knowledge component. This chapter is rounded off (Section 3.6) with a brief summary.

3.2 A DSS as an integrated software system

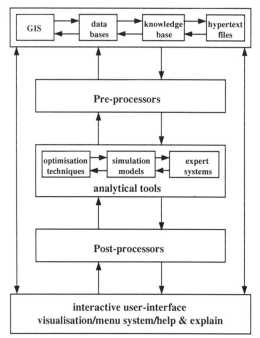

Figure 3.1 WaterWare system architecture (Jamieson and Fedra, 1996).

In the last couple of years development of DSSs has become a reality.[1] Many of developed DSSs are operational, although still not widely used. Important work on DSS content and structure has been done, being a response to demands from practice, rather than implementation of DSS theory. Advances in IT created possibilities for integration of large numbers of software tools and development of quite complex DSSs.[2] DSS complexity could be handled only by imposing a proper structure. Although structure depends considerably on content of DSS (and the content is again based on a DSS purpose), rather similar structures are proposed by various authors. Figure 3.1 shows the structure (or system architecture) of WaterWare - a DSS for river-basin planning (Jamieson and Fedra, 1996).

The purpose of developed DSSs varies substantially, but the tools (or types of tools) integrated in DSSs are often similar, if not the same. Hence an attempt has been made to create a 'Standard DSS Framework' (LWI,1998), a sort of generic content and structure of a DSS for integrated water management (Figure 3.2). Three main components can be distinguished in this generic DSS structure:

---

[1] No information on DSSs developed in the Netherlands could be found in 1996 to be included in the overview presented in Chapter 2. At workshop held in Wageningen (The Netherlands) just two years later, five operational DSSs were demonstrated. (However, no published information on demonstrated DSSs is available at this moment.)

[2] Some integrated software tools, i.e. DSS components, are complex even without integration (e.g. GISs, databases).

- the shell (user interface);
- the simulation models, and
- databases.

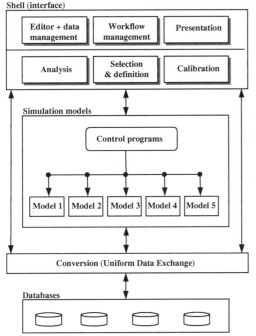

Figure 3.2 Standard DSS Framework (LWI, 1998)

The same components can be recognised in the majority of developed DSSs, regardless of their purpose and/or complexity (Figure 3.1 and Figure 3.3). Furthermore, the central place in a DSS structure is, almost by rule, occupied by simulation models. This fact confirms continuous appreciation for models in practice, regardless of their shortcomings. Moreover, the tools assembled in the DSS shell (Figure 3.2) are regarded as 'Generic Tools' because they are not 'model-specific and can be widely used'. Generic Tools are meant to guide and support workflow in the DSS, that is basically model simulation using data stored in databases. The survey of Generic Tools (LWI, 1998) ranged from spreadsheets to shell-development tools. Classification of the Generic Tools (as shown in Figure 3.2) appears to be quite rigid, because the majority of surveyed tools are multi-functional (for example, able to analyse, as well as to present results of the analysis). Nevertheless, the survey revealed some advances in the development of 'generic DSS tools' in real sense of the word. These are DSS development tools, or the tools meant to support this process. In one or the other way, they provide development of a DSS shell and integration of various modules (DSS components). The emphasis (with respect to tools features) is made on a model pre- and postprocessing (e.g. Stela, developer: HR Wallingford; CMT, developer: Delft Hydraulics), visualisation (e.g. AVS/Expres, developer: AVS) and transfer of information among the modules (e.g. Infoworks, developer: Wallingford Software).

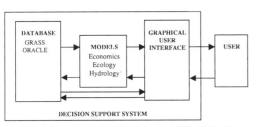

Figure 3.3 The components of the NELUP DSS (Dunn et al, 1996)

Further development of generic DSS tools can be expected that will lead to standardised, fully modular and flexible structures of DSSs. Notably, however, no separate place in Standard DSS framework (Figure 3.2) is left for any kind of knowledge component. It can be assumed that

Generic Tools 'Workflow management' and 'Selection & definition' should contain some kind of protocol and supporting information (on DSS tasks and sub-tasks). Unlike other potential DSS components (e.g. models and databases) which require only integration, these knowledge-containing tools need to be defined before anything else. Moreover, protocols are just a fraction of knowledge that could (and should) be encapsulated in the DSS.

Without knowledge components, DSSs remain simply integrated software systems meant for numerical information handling. As the influx of numerical information continuously grows, its handling becomes more and more complex. Encapsulation of complementary qualitative information in the DSS can alleviate handling of numerical information and increase DSS efficiency and reliability. This could be especially beneficial for simulation models (the core of a DSS) as shown by K. Fedra almost a decade ago (see previous chapter) as well as in his recent work (Figure 3.1). Systems like WaterWare (that contain encapsulated knowledge) belong to the so-called fifth generation of hydroinformatics systems (Abbott,1991). Although having relatively modest knowledge component, they represent 'comprehensive DSSs' or integrated information systems.

3.3 A DSS as an integrated information system

As information plays a central role in the decision-making process (see Chapter 2.1), it should be used as a main criterion in the definition of DSS structure. Accordingly, a DSS should be seen as a system that integrates information, rather than software tools. In principle, the support that a DSS is able to provide is proportional to the (qualitative and quantitative) information contained in the system.

Figure 3.4 Notion of integrated information

Partitioning of information on numerical (quantitative) and knowledge (qualitative) is essentially relative, because in a broader sense information as a whole (i.e. qualitative and quantitative) can be seen as knowledge.[3] This partition is, however, made to highlight knowledge that is not (or cannot be) expressed in digits, and processed and presented as such, as exemplified in Figure 3.4. Quantitative information is, for instance, required on:

– processes (represented by data, i.e. parameters and variables),
– processing tools (i.e. encoded techniques for numerical simulation/ modelling of processes), and/or
– procedures that employ data and processing tools for a purpose of problem solving.

---

[3] The term 'information' has its root in the Latin word *informatio*, meaning notion, understanding, teaching, introducing, announcing, reporting, etc. According to the Oxford dictionary, information is, in the English language, a synonym for knowledge, news, something told, an instance of this, etc. In practice, information is also often used as a synonym for 'data'.

The vast majority of contemporary DSSs integrate only numerical information (Figure 3.5); data storage, processing and presentation are usually the main DSS operations carried out by (mostly in this order) databases, models and GISs. Most processing (particularly simulation) is done by

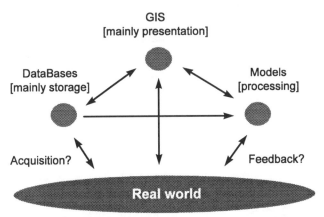

Figure 3.5 Integrated information: state-of-the art

models, although it is also performed by GISs and databases (for the purpose of data analysis); similarly, modelcodes and databases are usually capable of data presentation, but most presentation is done by GISs. GISs are in principle not meant for storage of large data quantities, nowadays often continuously required by the system. Therefore integration of powerful rational databases (e.g. Oracle) with models and GISs is

an imperative; this appears to be quite a complex, but still an accomplishable task (TNO, 1994; Sokol, 1996). The DSS is usually operated through the User Interface (UI), developed as a DSS shell, or as an extension of the GIS interface. User protocol (built in the UI) controls communication among DSS components, but communication with the 'real world' (Figure 3.5), especially in terms of acquisition and feedback, is still done individually. A DSS (as a system) should, however, have a unique input (of 'raw' information) and output (of 'processed' information or a 'new piece' of information).

Storage, processing and presentation of information discussed above are some of the main DSS information-related operations, but just some. Once processed and presented, information needs

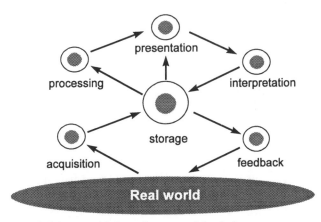

Figure 3.6 Integrated Information: state-of-the-design

to be interpreted. As with storage, processing and presentation, interpretation should take place in DSS, followed by storage of the results of interpretation (Figure 3.6); then the results can be used ('feedback' to the real world) or reused by the system. Finally, none of information-related operations can be performed if information is not acquired. Acquisition and Feedback are 'edge' (boundary) activities of a DSS. A DSS cannot perform acquisition or control use of results. Still it can provide support.

Carrying out any of the main DSS operations involves (and asks for) both, numerical information and knowledge (Figure 3.6). A DSS should be structured in such a way as to provide an 'optimal' use of available information, and that can be done only through a proper design and integration of DSS components. It is clear that the main operations must be carried out in a certain order; for instance, interpretation cannot be done before presentation, processing should be followed by presentation, etc. (this holds also for iterative procedures). DSSs are, however, developed to support one or more tasks, defined by the purpose of the DSS. Carrying out the tasks means carrying out one or (usually) more information-related operations. Task-oriented Knowledge-Based Modules

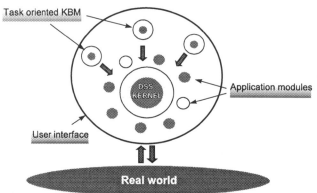

Figure 3.7 A DSS for GWQM - system architecture

(KBMs) could therefore be designed to support the user in performing the tasks (Figure 3.7). Content and design of KBMs are strongly dependent on tasks they support, but in principle, they should provide information on operations required for task accomplishment: which operations are required, why, how to perform them and when (i.e. in which order).

A DSS needs to have a common (single) storage (DSS kernel - Figure 3.7), at least as far as numerical information is concerned; all the data coming from, or going to the real world should eventually be stored in a DSS kernel. It seems that the best way of developing a DSS kernel is to integrate a powerful rational database and a GIS. This is done by REGIS (TNO,1994), resulting in integration of spatially- and temporally-distributed data in a unique 'data model'. The other main REGIS constituent is a so-called 'application environment' that consists of modules that perform information-related operations (mostly processing and presentation). Some of these modules (Figure 3.7) are build as an extension of a 'data model' providing a specific presentation (e.g. geohydrological columns) and/or processing (e.g. statistics for geochemistry). The other modules can be considered as external, being software tools (not developed for DSS but) just integrated with REGIS (e.g. groundwater models).   REGIS is a (REgional Geohydrological) Information System that integrates numerical information in a comprehensive and effective manner. The system is conceived (conceptualised) as a task-oriented system, as confirmed by looking at the internal modules, each of which was developed for a particular task. Nevertheless, the tasks that REGIS should perform are not clearly defined as yet, and consequently, no protocol-based UI is developed so far.

REGIS integrates information, but does not provide support (in DSS terms). Task-oriented KBMs are needed to 'transform' REGIS (or a REGIS-like) system into a DSS. That brings back questions of knowledge encapsulation, knowledge processing and the DSS knowledge component. These questions are discussed in Section 3.5. In continuation, (the framework of) the DSS for Groundwater Pollution Assessment is introduced, being implementation of the concept of an integrated information system.

3.4 The DSS for Groundwater Pollution Assessment - the framework

The DSS for Groundwater Pollution Assessment is conceptualised as a integrated information system, meant to support assessment of point-source groundwater pollution problems. The analysis of the DSS application realm in groundwater quality management (Chapter 2) is used as a basis for defining purpose, and subsequently, the content of the DSS. Site characterisation, vulnerability assessment and groundwater modelling are acknowledged as the main tasks that need to be carried out while assessing groundwater pollution at the local scale. Accordingly, three task-orientated Knowledge-Based Modules (KBM) were developed to support these tasks (Figure 3.8). These KBMs make up the DSS knowledge component. General considerations related to development of the knowledge component are given in the next section. KBMs for Site Characterisation (SCM), Vulnerability Assessment (VAM) and Groundwater Pollution Modelling (GMM) are described in detail in chapters 4, 5 and 6, respectively. The intention is to extend the knowledge component with a Case Based Reasoner (CBR); the role of a CBR in the DSS is outlined in Chapter 6, using modelling case studies as an example.

The DSS kernel is REGIS, a geohydrological information system, having a GIS (Smallworld, Arc-Info or Arc View) and a rational database (ORACLE) as its main components. REGIS provides a unique possibility for storage of both raw hydro(geological) data and data interpretation. That means that hydrogeological maps (or rather the hydrogeological model) available from REGIS are a result of hydrogeological interpretation (made by experts), rather than an outcome of automatic interpolation. The case studies (whose re(use) will be controlled by the CBR) are also to be stored in REGIS; REGIS (DSS kernel) would, in that way, become the central DSS information storage.

Only a few application modules are integrated in the DSS to date, mainly because of the time consuming development of KBMs being the prime task of this research project. Integration is made with the PMWIN modelling environment (Chiang and Kinzelbach, 1996), which is a pre- and post-processing shell around a few groundwater modelcodes (including MODFLOW). Connection is also made with the REMOD module, which (like PMWIN) uses REGIS output files to prepare MODFLOW input files (Chapter 6). Unlike PMWIN, REMOD does not provide any pre- and/or post-processing support. REMOD is developed as a part of this project, and it is meant for the modellers who do not use commercial (PMWIN-like) shells. The integration also included development of options in REGIS for preparation of a MODFLOW grid and spatially distributed grid-like data files that are used as an input by PMWIN and REMOD.

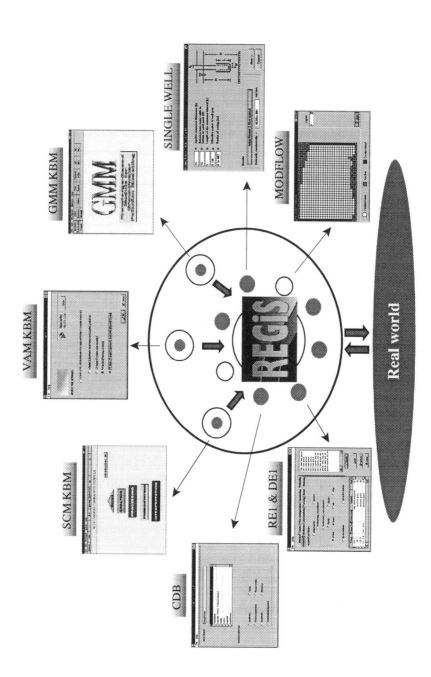

Figure 3.8 DSS for Groundwater Pollution Assessment

The DSS integrates a Chemical DataBase (CDB), that is used by all KBMs (Figure 8). The CDB is the 'Soil Transport and Fate Database' (Sims et al,1991) originally developed for MS DOS. The MS Windows (NT 3.51) version was developed as a part of this research project.

As an example of SCM application, the 'Single Well Solution Software' (StreamLine Groundwater Applications, 1996) was used (Chapter 4).

The DSS also integrates the 'RE1' and 'DE1', modules that are developed for estimation of retardation factor and dispersivity parameter (respectively). Although described as a part of GMM (Chapter 6), RE1 and DE1 are stand-alone applications (see next section and Chapter 6) and can be considered as a part of the application environment.

At this stage of the DSS development, no User Interface (as given in Figure 3.7) was created. The UI should contain the protocol composed of the main assessment steps, where special attention should be paid to various interactions among DSS components. For instance, results of site characterisation are used by VAM and GMM, the DE1 module can be triggered by any of KBMs and REGIS is required for each of the main assessment tasks. Nevertheless, further work on specification and systematisation of DSS tasks (and subsequent improvement of KBMs) is needed before a unique DSS UI is developed.

The DSS for Groundwater Pollution Assessment is far from being complete and/or operational. The main purpose of this research project was, however, knowledge encapsulation, rather than development of operational software; and unlike most of the DSSs, this one contains a substantial knowledge component, further discussed in next section.

3.5 DSS Knowledge component

The knowledge component is an integral part of a comprehensive DSS. It can be composed of task-oriented KBMs, that, in return, can contain knowledge represented in various forms. Nevertheless, design and development of a knowledge component are the last steps of the knowledge encapsulation process. Knowledge encapsulation is often understood as transformation of the knowledge into an electronic form and its storage in the computer. The encapsulation is, however, a much more complex process that includes:

- Definition of the problem;
- Knowledge acquisition;
- Knowledge systematisation;
- Knowledge formalisation;
- Design and development of a knowledge component;

– (Integration of the knowledge component into a DSS).[4]

First three steps can be quite difficult to carry out, especially if the problem is case-specific, if the problem-solving asks for interdisciplinary knowledge and if knowledge on the problem is not complete. Additionally, it might happen that standard knowledge representation forms cannot be applied (step 4), due to the listed obstacles, or simply because not all the knowledge can be represented by rules, decision tables, semantic networks or frames. In a such situation, that very much holds for groundwater pollution problems (Chapter 2), possibilities for development of a DSS knowledge component seem quite limited. Possible ways to circumvent (overcome) the obstacles and encapsulate knowledge on groundwater pollution problems are suggested in the following three chapters.

Problem definition, the first step of the knowledge encapsulation process is very often underestimated. A problem is defined by means of objectives and the main tasks that should be carried out in order to fulfil those objectives, i.e to solve the problem. The poor taxonomy of the GWQM tasks has already been recognised during the search for a possible DSS application realm (Chapter 2). Groundwater reports (case studies) contain, quite often, fairly precise descriptions

---

[4] Some terminological clarifications are necessary at this point. With the rapid enlargement of electronically encapsulated knowledge, new techniques have been developed to assist (re)structuring and more efficient re(use) of this knowledge. Moreover, whole new disciplines have emerged, bringing along new, not always clear, terminologies. *Knowledge technology* is seen, for example, as a direction within IT having as its main tasks to:

– get insight into knowledge of complex decision-making processes, and
– automate complex decision-making processes into a knowledge system.

The former task is basically about knowledge acquisition and systematisation, and the latter about formalisation and design and development. Thus knowledge technology is about the encapsulation process. *Knowledge modelling* can be considered as a supplementary step of the encapsulation process, being development of 'standard models' or templates for various decision-making processes (classifying, configuring, repairing, predicting, etc). KADS (Knowledge-based Approach to System Development) methodology (e.g. Wielinga et al, 1993) is used to 'model' actual knowledge to fit one of the standard models (expert, organisation, communication model, etc.).

Knowledge technology, as defined above, is sometimes classified as (just) one of the IT technologies used in *knowledge management (KM)*. KM could be defined as 'the systematic process of finding, selecting, organising, distilling and presenting information in a way that improves an employee's comprehension in a specific area of interest'. A distinction is made between KM, which establishes 'a direction the process should take', and *knowledge engineering*, which 'develops the means to accomplish that direction'. Knowledge encapsulation, as defined above, reflects both KM and knowledge engineering. KM is a wider term, because it also deals with knowledge that does not necessarily have to be electronically encapsulated, and with the knowledge processing. Besides, KM takes an organisation (the user is 'an employee') as a prime domain, and not a particular (management) problem (as encapsulation does). In practice, KM is mostly about setting up the right framework to deal with knowledge in organisation (knowledge infrastructure) and about the processing of already encapsulated knowledge. Processing is (usually quite rapidly) passed to knowledge engineering, and carried out by developing and implementing various IT technologies. In this terminological fuzziness, it seems that knowledge engineering partially takes care of knowledge encapsulation (by developing/using knowledge technology (latter task - see above)), as well as processing of encapsulated knowledge (by developing/using other IT technologies). More on these 'other' IT technologies (groupware technology, intranet/extranet technology, knowledge discovery, etc.) will be discussed later in this section.

of objectives and related tasks. These are, however, difficult to find back in general protocols, guidelines and manuals. As expected, the most favourable situation is, relating this issue, in the field of groundwater modelling. That is due to extensive use of modelling (and work done on its improvement), but also due to its deterministic nature since a substantial part of modelling process is precisely defined numerical procedures. Problem definition for vulnerability assessment and especially for site characterisation appears to be more difficult than for modelling. It can be argued that only substantial narrowing of the scope can allow feasible generalisation of the objectives and related tasks. In other words, a certain (narrow) type of the problem should be selected (e.g. point-source pollution problems in intergranular medium, caused by particular pollutants, and so on...) in order to enable precise definition of the problem in terms of objectives and tasks. That is certainly a viable argument, but the question is how far to go in narrowing the scope; eventually, one would end up with development of DSS knowledge component for a single (case-specific) problem! The answer is probably (as usual) somewhere in the middle; the scope should be kept as wide as possible (to accommodate a variety of case-specific problems), but still narrow enough to reflect most of the characteristics (i.e. 'common' characteristics) of case-specific problems. Handling of the specifics of the real world could be alleviated by developing flexible and adaptive DSS components. Besides, various techniques developed for the processing of (encapsulated) information (see below) can be incorporated in a DSS to deal with the specifics of groundwater pollution problems.

Problem definition is crucial, because it determines not only the scope of the DSS knowledge component, but also the purpose and content of the DSS as a whole. Therefore it asks for teamwork between managers and specialists from various fields because of the interdisciplinary character of groundwater problems. The same holds for knowledge acquisition and systematisation. These two steps are partially overlapping, because some (at least preliminary) structuring is made during the acquisition, and that is used as a basis for the further acquisition (see Chapter 6). It seems that available AI knowledge acquisition techniques and procedures are suitable primarily for acquisition of knowledge from narrow, well-defined knowledge domains. Subsequent systematisation and formalisation of knowledge coming from these domains lead to application of standard knowledge representation forms. Contemporary DSSs can, however, accommodate knowledge encapsulated in various ways (e.g. hypertext-based software - see below). Besides, acquisition of large quantities of semi-structured, interdisciplinary knowledge asks not for 'classical' interviews with experts (classical acquisition), but for their intensive and interactive cooperation. Probably the most important postulate of that cooperation is 'consensus'; unfortunately, it often has to be reached about issues that are considered as a 'common knowledge' (thus not about 'frontiers of the known' i.e. current research issues).

How 'uncommon' common knowledge is, illustrates the fact that terms like 'aquifer' and 'groundwater system' still mean different things to different people. Cooperation in practice is not sufficient, not only in writing, for example, the books (which always incline to have a personal stamp), but also in preparing documents like protocols, manuals or guidelines. These documents ask for intensive cooperation, because they are supposed to specify a precise course of action, or at least actively to support it. It is striking that the demand placed on knowledge encapsulated in paper-documents becomes much higher after its 'transformation' into electronic form. The user of electronically-encapsulated knowledge expects direct guidance and/or active support. Software (and knowledge) engineering provide the means to increase the effectiveness of knowledge use (search mechanisms, knowledge processing), but user expectations can only be met if problem definition, acquisition and systematisation are carried out in the right manner.

A very pragmatic approach to knowledge encapsulation should be adopted; there is always more knowledge to be acquired and, consequently, systematisation can always be improved. Encapsulation is a continuous process, meaning that a DSS component should be designed and developed in such a way as to enable modification (improvement) and augmentation of encapsulated knowledge (a 'learning' system). Development of 'flexible' software has certain technical limitations that are inversely proportional to the quality of the originally encapsulated knowledge. In other words, if the latter was poor, and as such used in software development, no substantial adaptations would be possible; the only way is development of a new software (version). The choice between encapsulation of relatively poor knowledge (definition of 'poor' is rather case-specific) and no encapsulation at all, depends basically on the effort needed for encapsulation and the possibilities for improving encapsulated knowledge. In principle, the latter choice is more favourable, merely because of the availability of various IT techniques for processing of encapsulated information (see below).

Explicit knowledge, fairly structured and formalised in rules, can be encoded in a 'classical' rule-based knowledge base. That has been done with knowledge on the complexity of groundwater modelling (Chapter 6). The MCM rule-based knowledge base was developed to assist the user in setting up a groundwater model. The main goal of MCM development was to provide transparency of encoded rules and the possibility for their modification. The MCM is part of GMM, the knowledge-based module for groundwater pollution modelling (Chapter 6).

Some groundwater problems can be fairly defined in terms of objectives, tasks and parameters. Knowledge acquisition shows, however, that not all relevant parameters can be precisely defined and/or quantified. Moreover, the relative contribution (importance) of individual parameters varies from situation to situation. Ranking methodologies are usually introduced to deal with those kinds of problems. Encapsulation of methodologies such as that for vulnerability assessment (Chapter 5) asks again for software adaptability. If the user is able to modify (improve) parameter description, disregard parameters or introduce new ones, software becomes interactive in the real sense of the term. These software features (options) are provided in VAM, the knowledge-based module for vulnerability assessment (Chapter 5).

Definition of the problem leads to establishment of a procedure or a protocol that can be used as the core of a user interface in a knowledge-based system. To perform a step in the procedure could mean to make decisions, such to choose parameter value (VAM module) or to provide a yes/no answer (MCM rule-based knowledge base). In these cases, decision-making seems like a quite straightforward procedure, but the quality of decisions made depends on the information that they are based upon. When selecting a parameter value, for example, the user needs to know what the parameter represents, why and when it is required, what are plausible ranges, which methods are (could be) used to define the parameter, what are the processes involved, etc. Software applications can, therefore, be developed to support parameter selection by providing (containing) information directly required for the selection. This is especially useful if parameter selection is complex and/or frequently demanded, as, for example, selection of dispersivity coefficients and retardation factor. The DE1 and RE1 modules were developed to support selection of these parameters (respectively) and described in Chapter 6. Although considered as a part of GMM, DE1 and RE1 are in fact stand-alone applications that can be independently used (Chapter 4 and 5).

Not all decisions are so straightforward as those on parameter selection. Complex decisions involve a number of interrelated factors, meaning that, at the same time, complex, interrelated knowledge needs to be taken into consideration. This knowledge can be immense, interdisciplinary, scattered among various (sometimes inconsistent) sources, often poorly structured and, therefore, difficult to encapsulate. At the current stage of IT development hypertext-based technology seems to be the best way to deal with this problem. It allows integration of large quantities of interdisciplinary, poorly structured information in electronic form. Knowledge contained in the SCM, the knowledge-based module for site characterisation (Chapter 4) and partially in GMM (Chapter 6) is encapsulated by using hypertext 'topics'; topics contain information presented in textual and graphical form, as well as links with other topics and/or procedures located elsewhere in the DSS. In most cases the quality and quantity of knowledge encapsulated in topics cannot be regarded as optimal. However, the main advantage of hypertext topics is that they can be modified easily and augmented. Moreover, topics can be used as a basis for further structuring and formalisation of encapsulated knowledge. Hints, advice and recommendations can be, for example, 'extracted' from the topics and 'transform' into rules. That is already subject of 'knowledge processing' and beyond the scope of DSS development.[5]

---

[5] Knowledge processing is a broad term, especially if it includes processing that does not necessarily produce a new piece of information. If that is the case, then all kinds of search performed (by a simple browser or sophisticated CBR) on encapsulated information is also knowledge processing. The search provides information that can assist the user to make a conclusion (inference), i.e. to produce a new piece of information. Conclusions can also be made by inference engines in rule-based knowledge bases, but they cannot be developed for all kinds of complex problems. The trend in IT is, therefore, to develop technologies that would assist the user in the decision-making process. Internet technology, virtual reality technology, document imaging technology, and similar, are intended to provide fast access to the right (combination of) pieces of information presented in the most adequate way. Much attention is also paid (through, for example, groupware technology) to electronic communication, cooperation and coordination, emphasising in that way the importance of teamwork.

Knowledge discovery should also be mentioned here, because it implies knowledge processing. Actually the term Knowledge Discovery from Databases (KDD) is more appropriate, because it is about discovering information from databases. The core of the KDD process is the application of specific data-mining methods for pattern discovery

(continued...)

Nevertheless, knowledge processing should be briefly addressed at the end of this section being interesting from the point of view of DSS content and design.

If a DSS is seen as a task-oriented system, then DSS knowledge components can be designed as a collection of task-oriented knowledge-based modules. The knowledge-based modules can contain knowledge encapsulated in various ways (see above). It has been stressed several times that a DSS knowledge component should be developed in such a way as to allow modification and augmentation of encapsulated knowledge. Once developed, the knowledge component should be considered as dynamic, i.e. as the component whose content is used, but also constantly improved. That is an opportunity for knowledge processing.

3.6 Summary

What has been the main progress made in DSS development in the last couple of years? This question was posed in the introductory section of Chapter 3. In practice, a DSS is still seen as an integrated software system (Section 3.2). Decision-making is, in principle, based on available information, therefore, a DSS should be regarded primarily as a integrated information system (Section 3.3). The principles of an integrated information system were applied in the development of a DSS for Groundwater Pollution Assessment, whose framework was presented in Section 3.4. In the development of this DSS, special attention was paid to the DSS knowledge component, i.e. the component that contains encapsulated knowledge (Section 3.5). The knowledge encapsulation process was outlined, pointing out the importance of particular encapsulation steps, and proposing various approaches to knowledge encapsulation.

---

[5](...continued)
and extraction (e.g. Fayyad et al, 1996). In terms of knowledge processing, much more interesting (than KDD) is, for example, Terminology Management, which is about electronic extraction of the terms and their organisation (e.g. Ahmad, 1995).

4. KNOWLEDGE-BASED MODULE FOR SITE CHARACTERISATION

Indocti discant et ament meminisse periti
(those who do not know - to learn, and
those who know - to find pleasure in reminding)

Henault, French poet and historian

4.1   Introduction

Characterisation of an investigated site is the first task to be carried out when dealing with point-source groundwater pollution problems. The main characterisation objectives are conceptualisation of the groundwater system and diagnosis of groundwater pollution. However, the 'Site Characterisation' is rarely conducted in practice as an independent procedure that consists of a set of clearly defined steps. The site is usually described by the use of several well-known methods that yield parameters necessary for further assessment and possible remediation. Regulations and recommendations issued by governmental agencies (most often related to monitoring) are followed to just the extent that fulfils legal obligations. Due to an unsystematic approach and use of limited information (on factors, parameters, methods, etc.), Site Characterisation (SC) is often conducted superficially and inefficiently, yielding poor and sometimes misleading results.

Groundwater pollution problems are by definition 'site-specific', combining the specifics of a groundwater system with those of pollutant and management practice. To develop the site characterisation protocol (guidance) that would be applicable for each combination of these specifics would be an unrealistic task. (On the other hand, development of specific protocols would not be very practical either, before general guidelines are set up.) Nevertheless, groundwater pollution problems have much in common, that can be denoted as the 'generics' of the problems. In SC, the generics can be recognised in posed objectives, tasks to be carried out, parameters to be defined, methods to be employed, etc. Therefore, information on these issues, gathered, organised and presented through a simple, general protocol, would speed up and enhance the SC.

Groundwater pollution problems are interdisciplinary and information required for their handling seems to be immense. Luckily, contemporary information technology and software engineering provide the means to store (encapsulate) and subsequently handle efficiently enormous quantities of information. Therefore, a piece of software could be developed that would contain common, general knowledge ('generics') on site characterisation of point-source groundwater pollution problems. The software should be adaptable, so that newly-acquired pieces of information, especially knowledge related to the specifics of groundwater problems, can be added to the already encapsulated information. By developing the software, information required for SC will be made available to the user in a systematic and integrated form. That would substantially reduce the chance of some (important) piece of information being left out (forgotten) while characterising a site. By consulting knowledge contained in the software, the tasks such as the selection of the (relevant) parameters and (appropriate) methods for specific site conditions could be carried out easier, faster and with increased reliability. Gradually, a set of specific protocols can be developed, and even automated. Eventually, the efficiency of the developed software can be considerably increased by integration with other software that store, process or present information on the characterisation of the site.

The prototype of the Site Characterisation Module (SCM) has been developed as part of the Decision Support System for Groundwater Pollution Assessment. The main objectives regarding the development of the SCM were:

– to gather (acquire) general information (i.e. common, public knowledge) that is relevant for SC;
– to organise acquired information in a simple, but efficient manner;
– to encapsulate acquired and systematised knowledge into the module, providing possibilities for adaptation and augmentation;
– to integrate the module into the DSS: a) providing use of other software during the site characterisation, and b) permitting use of results of the Site Characterisation by other modules integrated into the DSS.

The posed objectives were fulfilled, but no assertions were made over (the most relevant) acquired knowledge, (the best) knowledge structure or (the best) knowledge representation. This was one of the first attempts to encapsulate and integrate actively this kind of knowledge (not just connect software) in the field of hydrogeology. Therefore, the development of the module would be justified if it succeeds to inspire (necessary) further development of knowledge-based modules.

The next section of this chapter (Section 4.2) contains the background information on: information required for SC, its sources and use; and possibilities to encapsulate knowledge on SC, and related previous work. The Module content and organisation is presented in Section 4.3, followed by description of software development and integration (Section 4.4). Two detailed examples of the Module's application are given in closing section of this chapter.

4.2 Background

SC is a complex task that includes gathering, processing and presentation (and interpretation) of information on a polluted site. Both quantitative and qualitative information is required for accomplishing of this task. Quantitative information is gathered: (1) by (single) sampling at the site; (2) from data bases (mostly results of continuous monitoring) and (3) from other sources (various maps, reports, etc). The SC processing is carried out at the site (when field methods are applied), in the laboratory and as part of a desk-top study. Presentation is mostly a desk-top task.

SC provides the diagnosis of a groundwater pollution situation. It is the independent task that precedes all the other tasks (related to the investigated site) such as assessment of the groundwater pollution situation, groundwater protection from pollution and/or remediation of polluted groundwater. Unlike the Assessment tasks (e.g. vulnerability/risk assessment and groundwater modelling) the SC does not (in principle) produce a quantitatively new piece of information (such as a risk score) and it is not directly related to prediction. The desk-top part of SC encompasses preparation of field and laboratory activities, and analysis, processing, presentation and interpretation of results obtained. The Site Characterisation Module (SCM) contains qualitative information, and it has been developed to assist a user in carrying out these tasks.

Quantitative information required for the SC could be stored (in databases), processed (e.g. statistically) and presented (e.g. by a Geographical Information System) within the DSS. Knowledge contained in the SCM is meant to assist the user in answering the question(s) 'which' (and 'why', 'when', and 'how') a piece of (acquired/retrieved, processed, presented) quantitative information has to be dealt with.

Knowledge can be, in very general terms, divided into general and specific knowledge (see Section 2.2.2). General knowledge is, in principle, common, public knowledge, while specific knowledge is less accessible. Specific knowledge could be used as a synonym for heuristic knowledge, if it could be obtained only from field experts ('expert' knowledge). Until recently, Artificial Intelligence (AI) has been striving to acquire and formalise primarily heuristic knowledge in rules, or more complex forms of knowledge representation (semantic networks, frames, objects). Contemporary AI strives, however, also to encapsulate much larger quantities of knowledge by introducing:

– Case-based reasoning. If the specifics of a problem (a case) cannot be (easily) extracted and represented by a set of rules, then the case as a whole can be encapsulated as a knowledge unit. Solution of a new problem (a new case) is found by searching, matching and adapting the most similar old case from the 'pile' of encapsulated cases (see Section 2.4.2).

– Hypertext-based software. HyperText Markup Language (HTML) was used to build the SCM. Therefore, some more attention will be paid below to hypertext-based software.

At first glance, the HTML looks too simple to be considered as a subject of Artificial Intelligence. However, the two following facts can be immediately brought up to eliminate such an impression: 1) the HTML is used for knowledge encapsulation; and 2) the broad application (and thus importance) of hypertext based-software is more than evident. Three kinds of HTML applications can be roughly distinguished: Internet, on-line help and converted documents.

The name '*Internet*' is self-explanatory; it connects an immense number of users in the world-wide network. Encapsulated information (knowledge) on Internet is given in the form of topics that are written in the and mutually linked. The topics contain text, images, tables, and links to other software applications. Internet content is completely accessible to the user: it can be viewed, printed, copied or downloaded. Much attention is paid nowadays to development of software that can be initiated and/or downloaded from Internet and subsequently run on various hardware platforms.[1] This is considered as a very important Internet feature, as is the user's possibility to augment Internet by adding a piece of his/her own information. Numerous useful pieces of information on groundwater pollution problems can be found on Internet sites. Those are usually overviews of activities of various companies and governmental organisations, occasionally accompanied by lists of references and downloadable software (e.g. Subsurface Remediation Information Center, US EPA). Contents, selected chapters, or even whole documents (recommendations, regulations) related to groundwater pollution are more and more often placed on Internet by governmental agencies or organisations (e.g. US RCRA - Resource Conservation and Recovery Act, and US CFR - Code of Federal Regulations).

Content-sensitive *on-line help* is a standard software feature written in the HTML. Help files are mostly used to explain a software application to which they are connected. Each software feature can be directly linked to a corresponding explanation contained in a help file. Like Internet, help files contain text, graphics and images; unlike Internet, they usually do not contain commands to initialise and run other software applications.[2] All contemporary groundwater modelling codes are supported by on-line help files. Besides a sole description of software commands, some of them contain basic information on model parameters.

The use of HTML to *convert documents* into electronic form is increasingly gaining popularity, especially since CD-ROMs became a standard hardware component. Software that contains encapsulated documents could provide numerous options for document browsing: contents, indexes, hypertext jumps, sequence browsing, hierarchical browsing, 'back' browsing etc. Some of these options are available in 'Surface Characterisation and Monitoring Techniques', a Desk Reference Guide (Eastern Research Group, 1993) that has recently been converted into an 'electronic book'. This document (i.e. CD-ROM) is integrated with the Site Characterisation Module.

Only two (valuable) examples of hypertext-based software that encapsulate general knowledge on groundwater pollution problems have been found so far. Both examples, namely a DSS for groundwater contaminant modelling ('OASIS') and a DSS for evaluating pump-and-treat remediation alternatives, were developed at Rice University, USA. The DSSs were described in

---

[1] There are various approaches to development of Internet client/server applications, such as Server Side Includes (SSI), Common Gate Interface (CGI) and Application Programming Interface (API). Development of sophisticated applications demands use of much more complex languages than HTML (C++, Perl,. Java, etc.).

[2] Exceptions are some 'standard' commands like 'print' and 'copy'. Contemporary help-developer tools, however, provide the possibility to build Macro commands (Macro programming language) to jump to another help file, execute a program, and so on.

Section 2.2.

The overview given above shows that some knowledge on groundwater pollution problems has already been electronically encapsulated in various forms. Encapsulated information on groundwater parameters, methods used for their determination and regulations that prescribe parameters determination, are very much needed for (or during) the Site Characterisation. However, no attempts have been made so far to develop software exclusively for Site Characterisation, a computer application that would assist a hydrogeologist in performing this task, thereby justifying the development of the SCM.

4.3   Module content and organisation

SC (or more precisely: its desk-top part) consists of several tasks, namely: Analysis, Processing, Presentation and Interpretation. Planning of (further) field investigations or 'Planning a campaign' could also be considered as a SC task. The SCM contains qualitative information (knowledge) that can assist the user in performing these tasks. The Module also includes commands that link qualitative information with procedures for handling (retrieval, processing, presentation) of corresponding quantitative (numerical) information. The majority of the procedures do not reside in the SCM, but in other software integrated into the DSS.

The SCM is organised in five TASK UNITS, as shown in Figure 4.1. SC begins with the ANALYSIS of a current problem; the user has to set up Objectives, to define Parameters required for fulfilment of Objectives and to choose Methods that are (or could be) applied in order to determine Parameters. Extensive information on Objectives, Parameters and Methods has been encapsulated in the Module. If field data are not available, the user will plan a CAMPAIGN. If data are available, data PROCESSING might be needed. The Module will explain what kind of processing is needed and why; moreover, the user could activate a processing procedure from the very same location in the module, if the procedure is available in the DSS. If processing has already been carried out (or if it is not needed), data have to be presented. The mode of PRESENTATION depends on objectives (e.g. combined presentation of parameters), parameters to be presented and, sometimes, on methods used for the parameter determination. Similar to the PROCESSING task, the Module will recommend the best way of presentation and provide the link with the related procedure. Systematic analysis, necessary processing and the most appropriate presentation make INTERPRETATION easier and more reliable. (Interpretation is closely related with reporting, that also takes place within the DSS.) If available data are insufficient for reliable interpretation, an additional Campaign should be planned. Otherwise, SC is complete.

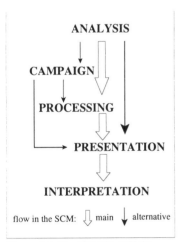

Figure 4.1 The SCM Task Units

Knowledge in the Module is organised in <u>Topic Units</u>, composed of sets of topics. Topics are hypertext-based knowledge representation forms that contain information presented in textual and graphical forms (images, diagrams and tables) and links with other related topics and other software. Each topic unit is dedicated to one of the main topics of the Site Characterisation. Task Unit ANALYSIS consists of four topic units: <u>Objectives</u>, <u>Parameters</u>, <u>Methods</u> and <u>Processes</u> (Figure 4.2). Other task units consists of a single topic unit and commands merged into the topic unit (Figure 4.2).

The protocol for SC is, at the same time, a basic flow in the Module. The flow in the SCM is shown in Figure 4.1 (task units only), Figure 4.2 (task and topic units), and finally in Figure 4.3. Figure 4.3 gives some more detail on the flow in the module, and depicts the connection with a subsequent DSS task (Vulnerability Assessment).

In the development of the SCM up to this point, attention has mostly been paid to the ANALYSIS Task Unit. The analysis is the most comprehensive part of the characterisation, forming the basis for all the other tasks; without thorough analysis, neither processing, nor presentation and (especially) interpretation make sense. Description of the ANALYSIS Task Unit will be given below, followed by some comments on the other task units and a brief discussion. Section 4.3 will be rounded off with a glossary of most important terms introduced and discussed in the section.

4.3.1 Task Unit ANALYSIS

During this, the very first step of the characterisation, the available data need to be analysed. The analysis should be carried out in the light of clearly specified objectives (Topic Unit <u>Objectives</u>). Fulfilment of an objective can be expressed in terms of determination of a parameter

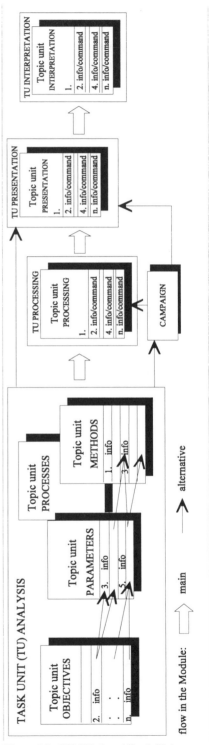

Figure 4.2 SCM Task and Topic Units

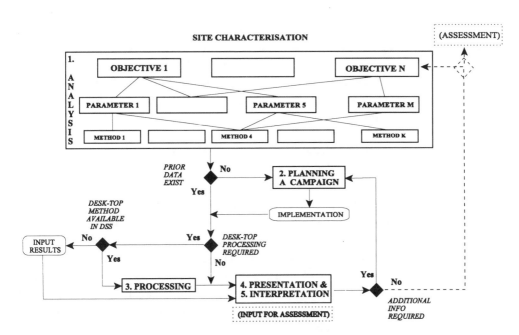

Figure 4.3 The SCM flowchart

set (Topic Unit <u>Parameters</u>). Eventually, methods (Topic Unit <u>Methods</u>) used for parameter determination have to be considered, particularly with regard to their accuracy and applicability. When the ANALYSIS is completed, the PROCESSING (if required) and/or PRESENTATION can take place. If data are not complete and/or not accurate enough, the ANALYSIS will reveal which parameters still have to be determined and which methods are the most appropriate for the purpose. Then the CAMPAIGN can be planned.

The topic units that comprise the ANALYSIS Task Unit (<u>Objectives</u>, <u>Parameters</u>, <u>Methods</u> and <u>Processes</u>) are interactively linked. The task is, however, carried out in the following order: objectives, parameters, methods. As indicated in Figures 4.2 and 4.3, more then one parameter can be required for fulfilment of a single objective (which is usually the case). Similarly, the same parameter could be required for fulfilment of different objectives. The same holds for the relation between parameters and methods; a parameter can (usually) be determined by various methods, whereas some methods determine (or contribute to the determination of) more than one parameter. The Topic Unit <u>Processes</u> is a 'background' unit, where information on various processes that control flow and transport of fluids in the subsurface can be found. This Topic Unit can be entered independently, as well as from various sites within the Task Unit, wherever explanation of processes is found necessary.

Topic Unit <u>Objectives</u>

The purpose and the content of Site Characterisation are defined by the objectives of characterisation. SC can be carried out at various levels of generalisation (from the preliminary to the very detailed study), but the main objective is always the same: to determine the state-of-the-art (diagnosis) of a groundwater pollution problem. The posed objective demands information on the contaminant (its characteristics and presence in the groundwater system) and on the groundwater system itself. Therefore, the main objective of SC can be divided into:

1. development of a conceptual model of the GroundWater System (GWS);
2. evaluation of groundwater pollution.

Definition of the nature and extent of soil and ground water contamination, as well as the potential migration of contaminants within the natural ground water system, requires basically collection and integration of geological, hydrological and chemical data. Development of a conceptual model of a GWS comprises collection and integration of information on: (1) geology: the physical framework within which subsurface fluids collect and flow; and (2) hydrology: the movement of fluids through this physical framework. The nature of the chemical constituents that are entrained in the system is defined through the fulfilment of the second objective of SC - evaluation of groundwater pollution.

To develop a conceptual model of a GWS involves defining the model components: system matrix, system boundary and boundary conditions, and system internal conditions (Figure 4.4a). The model components (see Glossary) are determined by the parameters of the conceptual model (Topic Unit <u>Parameters</u>).[3]

Evaluation of groundwater pollution comprises the (chemical) characterisation of pollution, and estimation of pollution extension (pollution delineation). This should, however, be preceded by estimation of basic groundwater (chemical) characteristics (Figure 4.4b). Two (sub)topics are attached to this objective of the Site Characterisation; one of them presents a 'two-stage approach' to the investigation of groundwater pollution problems (see Discussion), while the other describes sources and types of groundwater pollution.[4]

---

[3] Among all parameters, determination of system boundaries has absolute priority. Extension of the GWS is, at the same time, an extension of the site under investigation. Unlike other GWS parameters, determination of boundaries depends (besides geological and hydrological site characteristics) on other factors, such as the location of potential pollution receptors-targets (nature resorts, pumping fields, crop fields, etc). The system boundaries should be determined, as much as possible, in concordance with an aquifer(s) natural boundaries. Complete concordance is, however, rarely possible (see Glossary: definition of GWS).

[4] More then 30 different sources of groundwater pollution are categorised and described. For each source, an overview of the usually associated pollutants is given as well. Extensive information on pollutants (bacteria, viruses, nitrogen, phosphorus, metals and organics) is presented in a separate topic. 'Sources and Types' contain

(continued...)

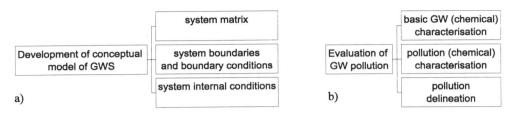

Figure 4.4 The main SC objectives and their components

Although SC is carried out within the boundaries of the GWS, some *regional study* is necessary in order to determine features that appear at local scale. 'Local' GWS (i.e. the investigated site) is usually a part of a regional GWS (with respect to lithology, stratigraphy, GW regime, etc). Therefore, an understanding of local features often requires prior understanding of their regional development. Additionally, a substantial piece of information on the groundwater system is transferred (interpolated) from the regional to the local scale. At the outset, boundaries of the local GWS cannot be determined without a regional framework. It was not found necessary to handle regional investigations as a separate objective. Instead, the parameters that should be analysed (processed, presented and interpreted) at the regional scale are indicated (Topic Unit Parameters) and linked with the Topic Unit Objectives.

Besides the two main ones, various additional objectives can be posed during the Site Characterisation. In the Module, they are considered as a 'third' objective:

3. additional characterisation

Additional characterisation may include description of managerial practice at and around the (potential) pollution source (liners, leachate collection systems, etc.), as well as specific pieces of information required for the following Assessment.

Topic Unit Parameters

Fulfilment of each (sub) objective requires determination of one or more parameters. Not all the parameters are of equal importance, nor is the importance of a single parameter necessarily constant. The user should be able to make a selection according to the specifics of the groundwater pollution problem at hand, information contained in the Module and personal experience.

---

[4](...continued)
'background' knowledge that is required for evaluation of groundwater pollution. A number of parameters (Topic Unit Parameters) also have links to these topics .

In the Module, the questions 'what?', 'why?' and 'when?' are answered by giving a definition of a parameter, followed by description of its purpose and importance. A determined attempt has been made to present basic information on parameters in a systematic and consistent way. Images, graphics and overview tables are used whenever possible, bearing in mind their functionality in conveying information to the user.

An overview of the parameters included in the Module is given below. This is (absolutely) not a definitive parameter list. The contents of the Module will be constantly updated (improved) by software developers, as well as by users (see Software Adaptability - Section 4.4.3).

1. The parameters related to the development of the conceptual GWS model are given in Figure 4.5.[5]

Rock types are described, especially with respect to their hydrogeological characteristics. This topic is also meant to be a reminder on basic classification and characteristics of rocks. Although at a first glimpse superfluous, the Topic is important, especially for the users with a non-geological background. Knowledge on rocks origin, genesis and composition is crucial for understanding aquifer evolution and nature. The nature and distribution of groundwater systems are controlled by lithology, stratigraphy (sedimentary rocks) and structure of the geological deposits. For the sedimentary unconsolidated rocks and soil, *texture* is an important parameter for lithological characterisation. *Colour* can be used as indicator of similarities or differences among the lithological units. Deposits of the sedimentary rocks of different age and lithology form so-called lithostratigraphic units. If the GWS contains lithostratigraphic units, their *spatial*

---

[5] Division of the conceptual model components and (subsequently) the model parameters is carried out according to the kind of information that should be obtained to develop a conceptual model, i.e. in accordance with the framework of the SC (see Objectives above).

Some geohydrologists would rather describe a GWS by the sets of input and output variables, state variables and system parameters. The difference between such a framework and one advocated in this report is not too large; system internal conditions coincide with state parameters (see Figure 5a), while system boundary conditions correspond with input and output variables (the exception is groundwater head, being state variable). However, this framework misses consideration of system unit(s). Besides, if 'state variables' are introduced, the basic division of objectives (i.e. on development of a conceptual model and evaluation of a pollution situation) would become meaningless (concentration, EC, pH, etc. are also state variables)!

A groundwater modeller would prefer to see 'hydrological stresses' or 'sources and sinks' as a separate set of parameters. In the modelling, internal sources and sinks (e.g. recharge and discharge wells) are not boundary conditions. This division is not always simple, for example in the case of net recharge which is sometimes a boundary condition, and sometimes a source term (for further discussion see Anderson and Woessner, 1992, p146).

Nevertheless, the content of the SCM is, as already stated, open to discussion.

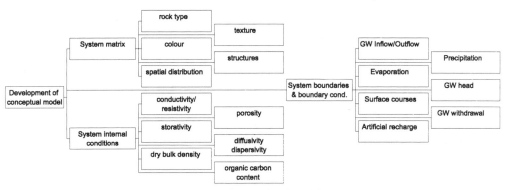

Figure 4.5 Parameters required for development of a conceptual model

distribution has to be determined.[6] Finally, various *structures* that occur in soil and rocks could substantially influence the spatial distribution of GWS units, as well as groundwater flow and contaminant transport within the system.

The importance of boundaries has already been stressed earlier in this chapter (see Footnote 3). Boundary conditions refer to the water inflow and outflow at GWS boundaries. If the GWS does not resemble an aquifer in its natural borders, a regional flow pattern has to be analysed in order to estimate its contribution to the system's *groundwater inflow and outflow. Evapotranspiration* is the most significant among processes that determine the difference between *precipitation* and net recharge. Net recharge influences the groundwater balance and *groundwater level* (groundwater head). *Surface water courses* (rivers, canals, ditches) can have a twofold function: to feed (recharge), as well as to empty (discharge) the GWS. The groundwater system could also be *artificially recharged;* this undertaking is often a compensation for the opposite process - *groundwater withdrawal.*

Groundwater flow and contaminant transport are determined by the properties of a lithostratigraphic (hydrogeological) unit, system boundary conditions, properties of contaminants and various mass transport and mass transfer processes (see Topic Unit <u>Processes</u>). The hydrogeological properties of the system units are described by several parameters, among which *permeability* (hydraulic conductivity) is the most important. Aquitards are described by *resistivity* rather than permeability. *Storativity* of the system is important, especially if seasonal oscillations cannot be disregarded. Effective *porosity* is required for estimation of actual velocity. *Diffusivity*

---

[6] Lithostratigraphic units are spatially distributed, having different hydrogeological characteristics. Field methods usually provide information on distribution of the units in one dimension. Point and linear information have to be interpolated (regionalised) in order to obtain the 3-D distribution of lithostratigraphic units. This is one of the two main steps in development of the conceptual model of a groundwater system. The second step is 'translation' of lithostratigraphic information into hydrogeological information. Not every unit is characterised as an aquifer, aquitard, aquiclude or aquifuge. A level of the characterisation (from preliminary up to very detailed) and (relative) lateral and vertical extension of the units, are (beside differences among units) the main factors that influence how the units are conceptualised. Accordingly, lithological units can be included as independent hydrogeological (groundwater system) units, or combined, or disregarded.

and *dispersivity* are the parameters that represent mass transport processes that take place in the GWS. Mass transfer processes are determined by the characteristics of a pollutant, as well as the hydrogeological environment. *Organic carbon content* and *dry bulk density* are the basic GWS parameters that are required for estimation of mass transfer processes (usually described by first order kinetics).

2. To fulfill the second (sub)objective of SC involves defining (firstly presence, and then) the properties and the extension of chemicals (pollutants) present in the GWS. The main parameters to be determined are listed in Figure 4.6, and briefly discussed below.

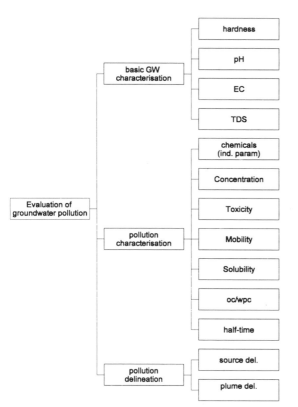

Firstly, *indicator parameters* have to be determined. That includes analysis of (soil and) groundwater samples on a selected set of chemicals that can, according to prior information (prior site investigations and general knowledge on types and sources of pollution) be suspected to act as pollutants within the GWS.

Indicator parameters are selected according to their prevalence, mobility and toxicity. *Concentration* of chemicals determined in analysed samples is the best indicator of their prevalence. The determined concentration has to be compared with the maximum concentration that is allowed in (drinking) water. Drinking water standards issued by governmental organisations do not allow the appearance of toxic

Figure 4.6 Parameters required to evaluate the pollution situation

chemicals or their appearance in concentrations that are considered to be toxic. *Toxicity*, therefore, has to be taken into account when selecting indicator parameters. Among the other factors, *mobility* (migration or transport) of pollutants through the GWS depends on the nature of the pollutants that are transported. The characteristics of pollutants such as the ability to dissolve in groundwater, to sorb with soil or sediments or to degrade in time, are basic for the chemical characterisation. Accordingly, at least the following parameters have to be estimated: solubility, organic carbon/water partition coefficient and first-order decay (half-life time). If the pollutant is soluble in water, its mass is transported with the groundwater flow. Besides its use for estimation of sorption potential, *solubility* can be used as a pessimistic upper limit of concentration. Together with the characteristics of the GWS, the *organic carbon/water potential coefficient* is needed for estimation of retardation due to sorption. Degradability could be caused by radioactive decay, biodegradation or hydrolysis and is usually expressed by the time required for the concentration to decrease to one-half of the initial concentration (*half-life* time).

Besides the test on potential pollutants, analyses of groundwater samples should include determination of the same basic groundwater characteristics such as *Hardness*, *pH*, *Specific electrical conductance* and *Total dissolved solids*. These parameters yield valuable information about groundwater constituents and contaminants.

According to the Two-Stage Approach (Connor, 1994) the chemical characterisation of the unsaturated zone (or the source zone) is followed by *source delineation*. Subsequently, the same procedure is repeated for the contaminant plume.

3 Parameters gathered under the third site characterisation objective 'Additional Characterisation' are primarily those meant as input for the Assessment, if such is planned (e.g. groundwater vulnerability/risk assessment or groundwater pollution modelling).

The knowledge-based module for Groundwater Vulnerability Assessment (VAM) uses a number of parameters already listed such as: hydraulic conductivity, hydraulic head, mobility, solubility, etc. Additionally, several parameters related to waste characteristics and managerial practice could be estimated during the Site Characterisation in order to accelerate the Vulnerability Assessment. If those parameters (e.g. liner, leachate collection, waste quantity, corrosiveness, etc) are not estimated in advance, they can be estimated from the VAM, but again via the SCM. The idea was to direct all the communication between 'raw data' (stored in databases) and the Assessment via the SCM - the module where knowledge on data is stored.

Topic Unit <u>Methods</u>

Methods are required for acquisition, processing and presentation of data (quantitative information). Some of the acquired (observed, measured) field data do not need processing, and could be directly regarded as parameters (e.g. groundwater head). Some data, however, require processing, i.e. some parameters are derived from the field data (e.g. an aquifer test).[7] Eventually, all the parameters (observed/measured or derived) have to be presented.

The methods can be roughly divided in three groups: field, laboratory and desk-top methods. The qualification 'roughly' is used, because the same method can be partially implemented at the site (acquisition), and partially in the laboratory and/or in the office (processing and presentation).

---

[7] If translation (interpolation) from point or line information to distributed (2-D or 3-D) information is also considered as processing (and not as presentation) then (almost) all the parameters are derived?

Laboratory methods have not been considered at this stage of SCM development. Attention has been paid to the field methods and the desk-top methods used to process and/or present the field information.

The document 'Subsurface Characterisation and Monitoring Techniques - A Desk Reference Guide' (Eastern Research Group, 1993) is the core of the Methods Topic Unit. This comprehensive Guide (approx. 900 pages) contains systematised information on more then 280 site characterisation and monitoring methods. Originally, the plan was to scan the document and integrate it into the Topic Unit, providing the connection with related parameters. The Guide has, in meantime, been converted in electronic form and become available on CD-ROM. It was therefore integrated in the SCM as such, considering the CD-ROM as an external (hardware) DSS component. The content of the Guide is briefly described below.

The methods described in the Guide are divided into: site characterisation methods (usually one-time sampling or measuring), monitoring methods (sampling or measuring over time) and field screening and analytical methods (mostly methods for chemical characterisation). There is, however, no clear dividing line between these methods. The Guide includes two volumes, of which, in general, the first one covers solids and ground water (sections 1-5), and the second covers the vadose zone (sections 6-10). The sections are titled as follows:

Section 1:   Remote Sensing and Surface Geophysical Methods;
Section 2:   Drilling and Solids Sampling Methods;
Section 3:   Geophysical Logging of Boreholes;
Section 4:   Aquifer Test Methods;
Section 5:   Groundwater Sampling Methods and Devices;
Section 6:   Vadose zone Hydrological Properties: Water State;
Section 7:   Vadose zone Hydrological Properties: Infiltration, Conductivity and Flux;
Section 8:   Vadose zone Water Budget Characterisation Methods;
Section 9:   Vadose zone Soil-Solute/Gas Sampling and Monitoring Methods;
Section 10:   Field Screening and Analytical Methods;

Each method described has an uniform summary that includes: methods description, method use, selection considerations, frequency of use, sources of additional information, etc.

If the data acquired by the field methods described in the Guide need to be subjected to further processing, the desk-top processing should, as much as possible, take place in the DSS. Additionally, complete parameter presentation can be carried out by using the DSS. The desk-top methods described in this Topic Unit are, actually, the procedures for data processing and presentation. All the site characterisation procedures available in the DSS (i.e. their short descriptions and the commands for their execution) are gathered in the task units PROCESSING and PRESENTATION.

Topic Unit <u>Processes</u>

This 'background' topic reviews processes that control migration and fate of contaminants in the GWS. Only the processes important for transport and transfer of miscible contaminants (contaminants dissolved in ground water) are considered. Transport of nonaqueous-phase liquids (NAPL) and multiphase flow in general, is left out at this stage of the Module development. Detailed information is provided on advection, diffusion, dispersion, sorption and degradation (radioactive decay, hydrolysis and biodegradation).

The processes are, in the same way as the features, being characterised by the parameters. Therefore, links with the relevant parameters are established while describing the processes. First the mass transport processes are presented, namely, advection, diffusion and dispersion.[8] These processes are described by the mechanical transport parameters (e.g. dispersion coefficient), being dependent on the characteristics of the GWS (Objective: Development of the Conceptual Model of the GWS). The impact of sorption and degradation on mass transfer is described by (bio)chemical parameters (e.g. the retardation factor), being dependent on the characteristics of the GWS, as well as characteristics of the pollutants (Objective: Evaluation of Groundwater Pollution). The links are also made with the procedures for derivation of complex parameters (e.g. the retardation factor).

The Topic Unit <u>Processes</u> was not included in the protocol (objectives-parameters-methods), assuming that information on processes might already be known to the user. However, this topic can be always used as a reminder. Besides, newly-acquired knowledge can easily be added to the topic.

4.3.2 Other Task Units

Information on Objectives, Parameters, Methods and Processes that is contained in the ANALYSIS Task Unit, provides the grounds for further acquisition, and/or processing, presentation and interpretation. Information that is required for carrying out a task (site characterisation, in this case) should, preferably, be integrated, systematised, formalised (as much

---

[8] The processes are divided in two groups (National Research Council, 1990): (1) MASS TRANSPORT PROCESSES (those responsible for material fluxes), i.e. advection, diffusion, dispersion; and (2) MASS TRANSFER PROCESSES (sources or sinks for the material), i.e. sorption, radioactive decay, dissolution/precipitation, acid/base reactions, complexation, hydrolysis, redox reactions and biodegradation. (All the processes from the second group are chemical mass transfer processes, except of biodegradation where the transfer of mass is being biologically mediated.)

The mass transfer processes, could be further divided into (Brusseau, 1994): (1) interphase mass transfers (such as sorption, liquid-liquid partitioning and volatilization) that involve the transfer of matter in response to gradients of chemical potential; and (2) reactions (biotransformation, radioactive decay, hydrolysis, etc) by which the physichemical nature of a contaminant is altered.

as possible) and presented in the context of the task.[9] Only then can the right decisions be made over further acquisition, processing and presentation. For instance, if the user (according to information available in the Module and personal experience) declares a parameter less important for the problem at hand, then: (1) a less accurate, but cheaper and/or faster method will be used to define the parameter; (2) the processing (if it takes place at all) will be simple, followed by (3) a less detailed presentation (e.g. the parameter will be presented with other parameters of similar importance), etc. Integrated information contained in the Module speeds up decision-making by providing substantial pieces of information (so demand for additional information search is limited) and by pointing out sources of additional information. In addition, it improves decision-making by not allowing an important piece of information to be forgotten (a virtual reminder).

Not only integrated information, but also the tools for information processing and presentation accelerate SC. Therefore PROCESSING and PRESENTATION task units, that gather information on (and connections with) procedures (for processing and presentation) are contained in the tools integrated in the DSS. The task units PROCESSING and PRESENTATION will be briefly described below. Prior to that, a (possible) role of the SCM in the planning of field investigations will be discussed (Task Unit CAMPAIGN). Eventually, some comments on interpretation of the site characterisation results will be given (Task Unit INTERPRETATION).

Task Unit CAMPAIGN

If no data are available, the planning of field investigations (planning a campaign) is the first step of the Site Characterisation. The campaign can be carried out in several stages, in which results obtained at previous stage determine the content of the next. If no data are available, the planning will proceed immediately after the analysis of objectives, parameters and methods. Otherwise, it might be needed to process, present and interpret data, before a decision on additional investigation is taken. Anyway, the superb graphical capabilities of Geographic Information Systems (GISs) could be utilised in planning a campaign. A GIS screen could therefore be 'transform' in a manner of an 'electronic notebook'. The contents of the screen would be defined by the specifics of a parameter under the investigation. For the majority of parameters that would be a background map with locations of previously- taken observations or measurements. The map could be accompanied by forms (specific for each parameter) for filling-in the field data. Besides, various tables, charts or diagrams containing information (on a parameter or a method), useful or required at the site could be added to the notebook. A 'sheet' of the notebook could be prepared for each parameter and briefly described in the Task Unit CAMPAIGN. The task unit should also include the commands for activating the notebook. Eventually, notebook prints should be taken along to the investigated site.

---

[9] 'In the context of the task' means that information should be presented (at least in general terms) according to the steps taken while conducting the task (i.e. according to the protocol). It would be quite difficult, at this moment, to develop more specific protocols, primarily due to the poor taxonomy of SC tasks and related knowledge. Development of the SCM is, however, a step towards it.

An example of the notebook for a campaign will be shown in Section 4.5.

Task Unit PROCESSING

The majority of the site characterisation desk-top methods are numerical procedures that can be encoded and integrated into the DSS. Some procedures have already been encoded, and their connectivity with other modules of the DSS is only matter of compatibility. Some of them can be improved by using GIS display options (e.g resistivity curve matching). Procedures such as those for calculation of hydrogeological parameters from aquifer test data and estimation of the amount of leachate from landfills, could become standard DSS options. In addition, a number of relatively simple procedures can be develop in order to alleviate site characterisation; an example is a procedure that compares measured chemical concentrations with standards retrieved from a chemical database. Additional examples are selection of an appropriate dispersivity coefficient, and selection of an appropriate equation for estimation of organic carbon/water partition coefficient (more discussion of examples is given in Section 4.5).

If interpolation of point to distributed information (2-D or 3-D) is considered as processing, rather then presentation, then the Task Unit PROCESSING should also contain procedures for producing a variety of maps required for the Site Characterisation. Interpolation is not a simple procedure, especially if more sophisticated methods are available (kriging, for example). Therefore experience with interpolation of various parameters should be added to the SCM.

Most of the procedures will be executed by databases and the GIS, so they would be (at least partially) written as queries and scripts (containing sets of commands for these software). The GIS already contains various procedures, such those for classification, overlay, or so-called neighbourhood operations, connectivity operations, etc. Parameter processing should make full use of these procedures.

Task Unit PRESENTATION

Each parameter requires a specific form of presentation; some simple examples are groundwater head (map-contour lines) and texture (a texture map and diagram-texture triangle). Parameter presentation is sometimes conditioned by the method used for parameter determination. Specifics of the problem at the hand can also influence presentation (e.g. which parameters should be presented together). Information on parameter presentation should be made available to the user together with the presentation procedures. The Task Unit PRESENTATION contains this information along with commands that link the Module with procedures. The procedures are executed by the GIS, which already contains predefined procedures for presentation of the same parameters. In a contemporary GIS, much freedom is left to the user in designing maps, charts, tables and other ways of presentation. No restriction on this freedom is needed, but the user should be offered a way of presentation which is (or should be - suggestion) commonly accepted as the most appropriate. That means, for instance, that a map should reference: coordinate system, orientation, author, data origin, date, etc; Practice teaches that these 'trivial' map attributes are very often 'forgotten'. The user can decide not to follow advise and not to use a pre-defined presentation procedure; however, that is then the user's own responsibility.

To conclude, a pre-defined form of presentation could be developed for each parameter that is required for SC. That would not only accelerate and improve SC, but also increase consistency in presentation. The importance of consistent presentation is explained below.

Task Unit INTERPRETATION

Systematic analysis, followed by the necessary processing and most appropriate presentation, make interpretation of the site characterisation easier and more reliable. Task-oriented analysis should be reflected in interpretation, as well as in subsequent reporting. That means that fulfilment of objectives is the central point of interpretation and reporting. Interpretation of a (developed) conceptual model of a GWS and an (evaluated) groundwater pollution situation should be carried out through interpretation of parameters. The report could contain an overview of the methods used, but the results of various method applications (that might be) important for a parameter determination, have to be gathered and interpreted together. Application of the SCM could substantially increase the consistency of the analysis, processing and especially presentation. Consistency alleviates not only the interpretation but also subsequent reporting. Reporting is of extreme importance for optimal use (subsequent assessment) and reuse (characterisation of similar sites) of the Site Characterisation results. If the reporting is consistent, the user becomes familiar with content of the report more quickly, meaning that less effort and time are needed to recognise the specifics of the problem. This ability is crucial for indexing of the case-studies (reports) that are also stored in the DSS.[10]

The Task Unit INTERPRETATION contains advice, hints and warnings related to interpretation of parameters. Through the set of commands, it should be linked with related 'chapters' of the Reporter, the other DSS component. Reporter is a text processor with a general reporting protocol that should be followed while writing the report.

4.3.3 Discussion

Experience gained during the development of the SCM prototype indicates that all the posed objectives were very demanding. The difficulties encountered related to knowledge acquisition and systematisation (i.e. the first two objectives) will be discussed here. (The Section 4.4 is dedicated to encapsulation and integration, i.e. the third and the fourth objective.)

---

[10] Indexes are used to retrieve the most appropriate case studies that might be useful for solving a new, similar problem. This is part of Case-Based Reasoning process (Section 2.4.2)

'What should be acquired ?', was the first and, at the same time, one of the most difficult questions. No overview of parameters that are required for the Site Characterisation could be found. Therefore the Desk Reference Guide was used to 'trace back' parameters that can be determined by the methods reviewed in the Guide. That has caused some disproportion among parameters gathered under the 'system matrix' of the conceptual GWS model.[11]

Very often information on the same issue was acquired from several sources, where none of them contained the complete information required, presented in the most adequate way; consequently the numerous pieces of information had to be extracted from various sources and combined. The various approaches to each issue, and the various styles and ways of presentation do not make this process easier. On top of that, the sources of information are not free of contradictory statements and definitions. The developer of the SCM has eventually to decide which statement or definition to include in the Module (therefore strengthening the call for cooperation in development of the SCM).

Various techniques have been developed for knowledge acquisition in the field of Artificial Intelligence. Most of them are related to extraction on narrow-domain, 'expert' knowledge. Nevertheless, attempts should be made towards broader utilisation of AI techniques (especially those related to case-based reasoning and terminology management).

Systematisation of the knowledge would be much easer if groundwater pollution problems were well-structured problems, so that steps taken in problem solving could be precisely defined. A very general protocol has been introduced at this stage of development of the SCM, dividing the task into only several steps: Analysis (objectives, parameters, methods), Acquisition, Processing, Presentation and Interpretation.[12] Work on the taxonomy of groundwater pollution problems (with respect to the site characterisation) and related knowledge is required before more specific protocols are implemented in the SCM. It should be noted that introduction of more specific characterisation steps (deepening) should be proportional to the freedom left to the user to make a choice (i.e. general steps - compulsory, specific steps - options).

---

[11] For instance, parameters 'spatial distribution' (of lithostratigraphic units) and 'colour' are treated equally, although the former is very complex (and relates to a dozen methods), while latter is simple and straightforwardly determined by one single method. However, no elegant way has been found (so far) to split 'rock type' and 'spatial variation' further into more parameters, without entering deeper into 'pure' geological knowledge. Complex geological knowledge is (indeed) sometimes required for full understanding of the content and distribution of lithostratigraphic units. Nevertheless, any attempt to continue knowledge encapsulation in this direction is seen (at this moment) as too big a challenge.

[12] Just one protocol has been acquired from the literature search: the 'Two Stage Approach' (Connor, 1994). This protocol has been only partially implemented in the SCM because it had been found too narrow (restricted); for example, it starts (immediately) with determination of indicator parameters. That does not mean that additional parts of the protocol should not be implemented in updated versions of the SCM. Therefore the 'Two Stage Approach has been completely scanned and made available (for use and discussion) as an independent topic within the Module.

More than 100 topics (entities within a topic unit) contribute to the structure of the SCM prototype. The fact that they are created just by following a simple, general protocol confirms the complexity of the problem. (It can be added that the number of links among the topics amounts to 500.) Despite the complexity, the first steps in developing the Knowledge-based Module for Site Characterisation have been made; some knowledge has been acquired, gathered and systematised (Section 4.3), and the prototype of the SCM has been developed and integrated in the DSS (Section 4.4).

4.4 Software development and integration

The Site Characterisation Module (SCM) is a hypertext-based software application, developed by using ForeHelp, a Help-Authoring System for Microsoft Windows (ForeFront Inc., 1994). Forehelp is, in other words, a tool for development of on-line help files. Accordingly, the SCM is (just) a file (run by the 'winhlp32' Microsoft Windows program), which however, does not act as 'ordinary' help file. Unlike help files, the SCM is not attached to any other piece of software in order to help in its use. On the contrary, other software is 'attached' to the SCM, meaning that in relation to that software the SCM plays a master role. One of the SCM functions (but only one of them) is to assist in use of the procedures that reside in the integrated software; the Module does that not only by providing information on the procedures, but also by activating those procedures. The SCM can therefore be considered not only as a DSS module, but also as a part of the DSS 'smart' interface.

The SCM was developed by using an on-line help editor rather than a WWW editor. [13] The choice was made according to the role of the DSS for Groundwater Pollution Assessment. The DSS is seen as a desk-top tool that is constantly used while managing groundwater problems (i.e. from data input to, and including, reporting). If the SCM was a 'loose' piece of software, just occasionally needed, then it could be placed on the Internet as a client/server application.

The major part of this section is dedicated to the knowledge representation in the SCM. Special attention is paid to hypertext, being crucial for knowledge representation, as well as for the flow in the module. This section also contains a description of some additional SCM features that were introduced primarily to enhance the flow in the module. Adaptability, considered to be very important for the SCM development, is discussed in a separate section. Eventually, the closing part of the section reviews the possibilities for the integration of the SCM with other DSS components.

---

[13] Tools for development of hypertext-based software are sometimes called simply HTML editors. There are some differences between HTML on-line help files and world-wide-web (WWW) files. The differences are, however, not substantial; HTML editors are already available that can not only 'edit' both types of files, but also convert one type to the other.

4.4.1. Hypertext and knowledge representation

None of 'classical' ways of knowledge representation (rules, decision tables, semantic networks, etc.) has been applied in the SCM, partly due to the sort of knowledge needing to be encapsulated, but mainly due to the stage-of-the-art of the taxonomy of knowledge in this field.[14] Basic knowledge representation is a 'topic'; topics contain text, images, graphics and tables. They are organised in Topic Units, and eventually in Task Units, as explained in the previous section. While structuring knowledge in topics, special attention was paid to definitions, purpose and importance of items (parameters, methods, procedures, etc.), using terminology, keywords, relations between items and referencing.

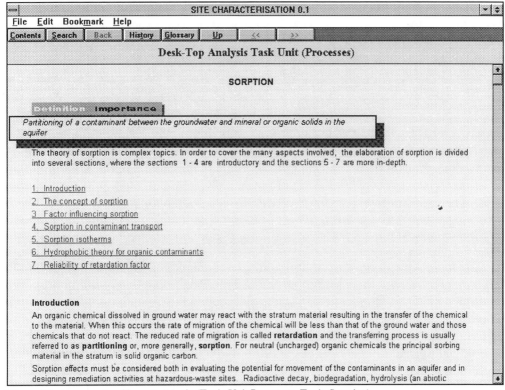

Figure 4.7 An example of hypertext topic (Topic Unit <u>Processes</u>, Topic Sorption)

To have basic knowledge on a site characterisation feature (e.g. system matrix) or process (e.g. sorption) implies being able to *define* it, and to identify its *purpose* and/or *importance*. In the

---

[14] Once relations between objectives (tasks), parameters and methods are clearly identified, rules, frames and even objects could be used for knowledge representation; the first step in that direction could be incorporation of decision rules in the protocol.

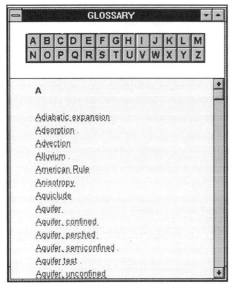

Figure 4.8 Topic Glossary

SCM, this knowledge is presented in several ways; if the knowledge is related to the topic as whole, then definitions, purpose and/or importance are given at the beginning of the topic; for example (Figure 4.7), a click on button 'Definition' opens a hypertext pop-up window that contain a definition ('pop-up' is a small window that appears on the top of window wherefrom the pop-up link is activated). Definitions of the terms used for the site characterisation features and processes are introduced to provide terminological clarity and consistency. All important *terms* (and their definitions) are gathered in the glossary (Figure 4.8).

Some other terms in the topic are also explained in the pop-up windows (but this time pop-ups are activated by clicking the highlighted text). This is the case if no additional information (besides the definition) is available in the Module. If there is, however, a topic that is dedicated to (or at least contains additional information on) a feature and process, the term is used to provide a link with that topic. The link can be established by browsing trough a keyword list or directly from the topic by triggering hypertext jumps ('jump' opens a new window and closes the one wherefrom the jump is triggered). Jumps are usually

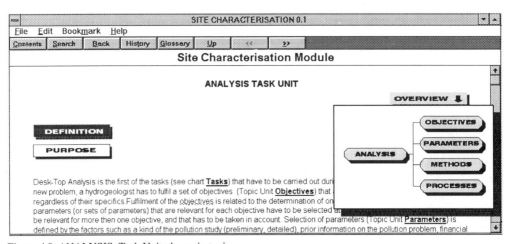

Figure 4.9 ANALYSIS Task Unit: the main topic

made to the top of the topic (i.e. top of the topic is set as a 'target'). However, there are exceptions (e.g. References, Processing Task Unit, etc), where a jump targets a particular reference, method or procedure located somewhere in the topic. Targets are extensively used for

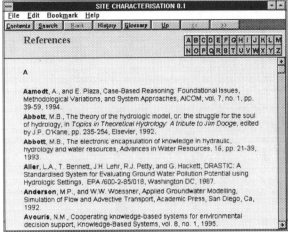
Figure 4.10 Topic References

quick movement within long topics; long topics begin with a contents, i.e. with the set of links to targeted sub-topics (Figure 4.7). These links, as made within a single topic, are called 'internal links'.

External links connect various topics in the Module. Topics usually contain more then one link in the form of highlighted text or 'hotspots' on images and graphics. In Figure 4.9, bold, underlined terms ('highlighted text') and the button 'Overview' contain links with the other topics. The topic that is linked can be the following protocol step, related item (topic), or a reference. References are sources of information used for the SCM development and additional sources of information; they are assembled in a separate topic 'References' (Figure 4.10). By analysing the content of the topic, the user learns which jump links with the next protocol step. The user might, however, also need some additional information (e.g. on processes) that is not directly related to the next characterisation step; other jumps contained in the topic can be used to open these related topics and acquire information (after which the user is supposed to return to the original topic). Once the content of the topic is mastered, it can serve as a reminder, bearing in mind that newly acquired knowledge can be added to the topic (see Software Adaptability). At that stage, the button 'Overview' can be used as a shortcut to the next steps. For example, by clicking the button on the topic Analysis (Figure 4.9), Topic Units of the Task Unit ANALYSIS will be listed; the overview of the Topic Unit Objectives provides links with the objectives, as well as related parameters (Figure 4.11). Overviews along with some other SCM features, are introduced to speed up flow in the module. The main module windows and their links are depicted in the Figure 4.12.

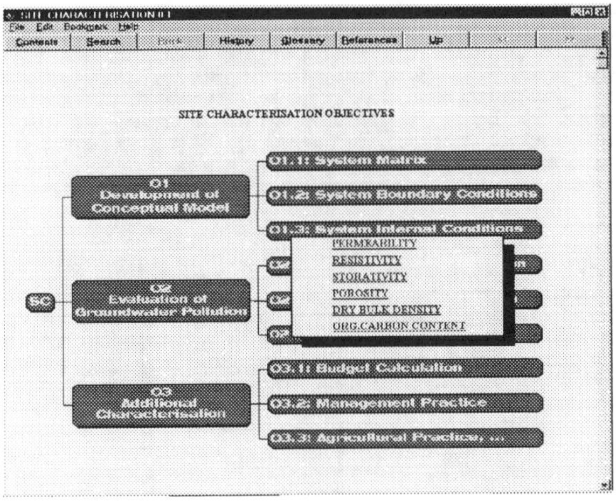

Figure 4.11 Topic Unit Objectives: Overview button

Instead of the 'Overview' button, individual parameter topics (Topic Unit Parameters) contain a 'Methods' button. Subsequent method topics (Topic Unit Methods) are supplied with

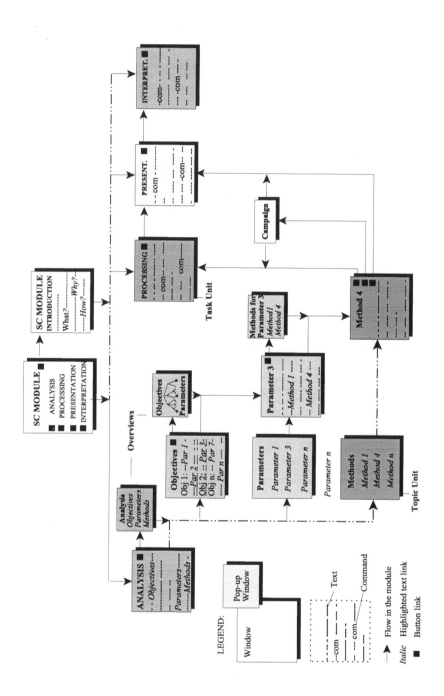

Figure 4.12 SCM: the main windows and links

'Campaign' and 'Processing' or 'Presentation' buttons. Eventually, the 'Processing' topic is linked to 'Presentation' and 'Presentation' with the 'Interpretation' topic. Highlighted text in these task units (see definition of task unit -Chapter 3) are the commands (macros) that link the SCM with the procedures for data processing, presentation and interpretation. Examples of the commands and procedures are given in Section 4.5.

4.4.2 Other SCM features

Besides the buttons placed in the topics ('topic buttons'), the SCM contains a collection of buttons that are constantly available at the top of a window (see Figure 4.7, for example). These 'module buttons' facilitate the flow in the SCM. The buttons are, together with some related module characteristics, briefly described below.

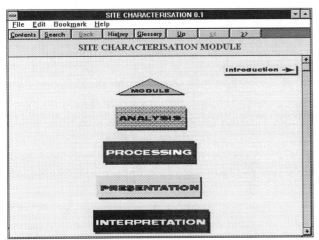

Figure 4.13   SCM contents window

The '*Contents*' button moves back to the main module window (or 'contents') from any position in the module (Figure 4.13). Besides four task unit buttons, the main module window contains buttons 'module' and 'introduction'. Button 'module' is linked to Figure 4.12. All the 'windows' (except the example windows for a parameter and a method) shown in the Figure 4.12 are 'hot spots', so the links can be established with the corresponding topics. Button 'introduction' opens a window where parts of this chapter are contained (Figure 4.15). Hypertext is again used to assist the user in becoming familiar with the Module. For example, button 'Buttons' opens a window where the module buttons are explained

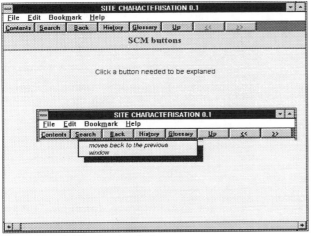

Figure 4.14   Self-explanatory buttons

(again by clicking the buttons - Figure 4.14). As this example shows, hypertext-based software is not only transparent, but also self-explanatory (description of a button function is obtained by pressing the button). That makes this kind of software very user-friendly.

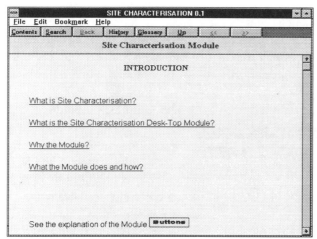

Figure 4.15   Introduction to the SCM

Button '*Search*' opens a standard Microsoft Windows search (Figure 4.16); by typing in a keyword, a list of related topics will appear in the lowest box. The keyword list is composed of the terms that are found important for the Site Characterisation. The selected terms are subsequently linked with the topics that contain information on (or involve, in one way or another) related site characterisation features and processes.

'*Back*' and '*History*' are also common on-line help features; the former returns to the previous window (previous step), while latter lists the steps previously made. '*Glossary*' (Figure 4.8) has been developed as a separate window class, i.e. it acts as a separate window application; once activated, Glossary will remain on screen as long as needed, regardless of actions (e.g. trigger of other module or topic buttons) taken in the SCM.

Button '*Up*' moves up to the previous level in the hierarchy (e.g. from 'System Matrix' objective to 'Conceptual model' objective, and further to 'Objectives' main unit topic, further to Analysis, etc).

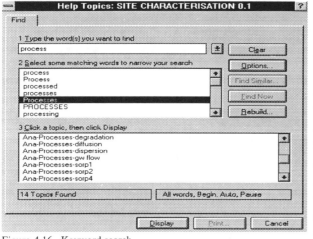

Figure 4.16   Keyword search

Hierarchy is established according to the site characterisation task that is pointed out by the protocol, i.e. according to the content of topic in the Module. Main hierarchic levels are shown in Figure 4.17. The level is high when the protocol points out a quite general task (e.g. Analysis) and the topic (e.g. main topic of Task Unit Analysis) contains 'general' knowledge (general description of the task). In contrast, the hierarchical level is low if the task and corresponding topic are relatively specific (e.g. dedicated to specifics of single parameter). Hierarchy has been introduced in order to structure encapsulated knowledge. Further work on this issue is required.

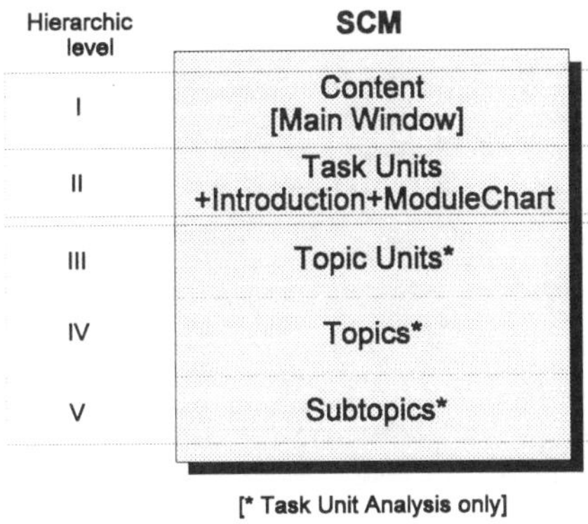

Hierarchic level	SCM
I	Content [Main Window]
II	Task Units +Introduction+ModuleChart
III	Topic Units*
IV	Topics*
V	Subtopics*

[* Task Unit Analysis only]

Figure 4.17   The SCM hierarchic levels

Browse buttons («») move to the next or previous window (forwards or backwards) in a browse sequence. Browse sequences link topics within the same hierarchical level. For instance, from the main topic of Topic Unit Parameters, the user can survey other topic units contained in Task Unit ANALYSIS. However, if a topic dedicated to one single parameter is displayed on the screen, then the browse buttons can be used to display all the other parameters in a sequence.

A hypertext-based software developer always creates a 'main menu' at the top of the screen (see Figure 4.7, for example). Among the main menu buttons, 'bookmark' can be mentioned while considering the possibilities for moving through the Module. Bookmark serves as reminder, but also provides quick moves to the previously marked places.

4.4.3 Adaptability

In order to develop fully operational software, the SCM prototype has to be improved with respect to content (encapsulated knowledge) and organisation. It is very important that any required software alterations can be carried out in a simple manner, so that a wide circle of developers can be involved. Besides, the SCM is seen as software that should be continuously augmented by the user; any new findings (new experience gained) on the Site Characterisation parameters, methods, procedures, references, etc. should be encapsulated in the SCM by the user.

If the same hypertext-based software developer (ForeHelp) is available, modification of the SCM is very simple; topics can be modify either directly in the ForeHelp developer, or in other environment and then retrieved in the ForeHelp. The simplest way to retrieve text files, rich text format (.rtf) files, graphics and images is through the windows clipboard (by 'copy' and 'paste').

ForeHelp, like other on-line help developing tools, produces topic (.rtf) file and so-called project (.hpj) file.[15] If ForeHelp is not available, the SCM (.rtf) file can be retrieved by any other HTML editor or even a text editor (but in that case some effort is needed to recover the original Module structure). Microsoft Help Workshop (a standard windows application) can be used to produce

---

[15] Rich Text Format (.rtf) is actually a hypertext topic file which can be generated in any HTML editor, and even in almost every text editor (Word, WordPerfect, etc). The project (.hpj) file is an ASCII text file that is used to compile a Help file. It contains all the information (e.g. on file location, screen options, macros, etc.) that the developer needs to combine topic (.rtf) files and other elements (e.g. other files, external bitmaps, etc.) into a Help file.

or modify the project file, and to compile the help file.

'Annotations', a standard on-line help option (located in the main menu, under 'edit'), is somehow also related to the adaptability. To 'annotate' is the simplest way of adding a new piece of information to the Module. This option is particularly attractive if the user wishes to preserve the original content of the SCM.

4.4.4 Integration

The Site Characterisation Module is an integral part of the DSS System for Groundwater Pollution Assessment. The integration provides straightforward use of other software required for the site characterisation, and the use of results of the Site Characterisation by other modules integrated into the DSS. The purpose of integration is to alleviate and to speed up the characterisation.

Software for data storage, processing and presentation is required for the desk-top part of the Site Characterisation. In the DSS, data are stored in REGIS (i.e. in the 'Oracle' database, that is part of the REGIS kernel). No special commands for data retrieval are introduced in the Module; however, basically every command for data processing and presentation implies data retrieval as a first step. Various options for data retrieval are available in REGIS; in some cases, the retrieval is (or can be) automated (by using queries or scrapbook, for example). Nevertheless, the range of the data set (spatial and temporal) to be retrieved has to be specified individually for each case-study (or even for each parameter).

Some data processing also takes place in REGIS (i.e. in the 'Smallword' GIS, that is part of the REGIS kernel). The GIS plays an important role in processing (classifications, statistics, etc), especially if interpolation is considered as a processing, rather than a presentation task. Besides REGIS, the SCM can be connected with software developed for data processing (e.g. aquifer test processing, geophysical data processing, etc); whether the software could be fully integrated in the DSS, or just loosely connected depends primarily on software compatibility. The SCM is already integrated with a few application developed to enhance characterisation, or to prepare input for subsequent Assessment. Some applications involve use of the Chemical DataBase (CDB), that is also an integral part of the DSS.

All the presentation in the DSS is done by REGIS (or via REGIS); REGIS options for presentation are almost limitless, so no additional software is required to handle presentation and output of the characterisation. Sufficiency of REGIS becomes even more evident since reporting also takes place in the REGIS module REPORTER.

In the section 4.3.1 (Topic Unit <u>Parameters</u>), use of SCM results in the following Assessment was addressed. The SCM is integrated with the knowledge-based module for Vulnerability Assessment (VAM), and the knowledge-based module for Groundwater Pollution Modelling

(GMM). Eventually, after discussing the SCM connections with the other DSS components, it should be recalled (Topic Unit <u>Methods</u>, Section 4.3) that the Module itself is an piece of integrated software; a part of the SCM resides on CD-ROM in the form of the converted document 'Subsurface Characterisation and Monitoring Techniques'.

4.5   Examples

Only two SCM applications are presented in this section, with the intention of explaining in detail how SCM applications are (and should be) designed and developed. Both examples are related to basic groundwater (quality) parameters (that have to be estimated during the Site Characterisation of groundwater pollution problems): hydraulic conductivity and solubility.

4.5.1 Hydraulic conductivity

The first example describes the role of the SCM in estimation of one of the elementary parameters: hydraulic conductivity. At the very beginning, the user becomes acquainted with the purpose of estimation of hydraulic conductivity in the framework of the site characterisation objectives (Topic Unit <u>Objectives</u>). Hydraulic conductivity is one of the key parameters that

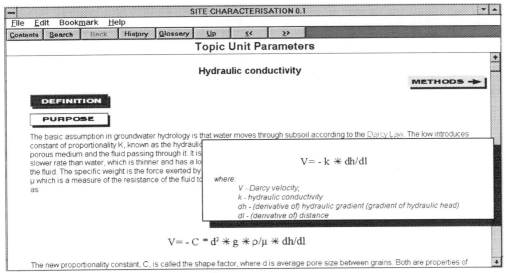

Figure 4.18   Topic Unit <u>Parameters</u>, an example

defines the 'internal conditions' of a groundwater system; the user learns that hydraulic conductivity is needed to define velocity of groundwater flow (and contaminant transport).[16] At

---

[16] At this point one of the 'complex' parameters is introduced: the groundwater velocity. Information is offered to the user that groundwater velocity can be derived by direct methods [1) multiwell observations of a naturally-occurring tracer, 2) multiwell tests conducted under natural gradient conditions with artificial tracers and 3) borehole dilution tests conducted with an artificial tracer (point-dilution tests)], or it can be determined indirectly

(continued...)

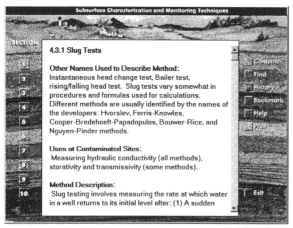

Figure 4.19 Desk Reference Guide, example a)

the same time, connection is made with another characterisation objective - delineation of groundwater pollution; velocity (that is to be estimated) is used in delineation of the present pollution plume (Site Characterisation), as well as for prediction of plume development (Assessment - groundwater modelling). In the next step, the Topic Unit Parameters will be activated (by clicking highlighted text 'hydraulic conductivity' or the button 'Overview'). Information on hydraulic conductivity will be displayed on the screen (Figure 4.18). If the user is familiar with content of the topic, he/she can proceed to the next topic unit - Methods (by clicking the button 'Methods'). This topic unit contains an overview of the methods for estimation of hydraulic conductivity. If conductivity data exist in the database, the user chooses the method (clicks the arrow) that is used for conductivity estimation. If no (or not enough) data are available, the user is advised to consult an overview table in Desk Reference Guide, by activating the icon 'CD-ROM'. The table presents the overview of aquifer test methods and contains links to topics where the methods are described. The topics are given in a consistent manner, including: other method names (synonyms), application, concise method description, method selection considerations (advantages and disadvantages), frequency of use, guidelines, and references. An example, 'Slug Tests' topic is given in Figure 4.19. The topics are linked with graphics - drawings,

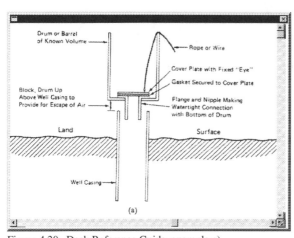

Figure 4.20 Desk Reference Guide, example a)

such as one showing apparatus for performing water injection slug tests (Figure 4.20). (A few animations are also included in the Guide!).

---

[16](...continued)
from the hydraulic conductivity, hydraulic gradient and effective porosity of the stratum material.

If results from the direct methods are available, or if the user decides to apply one of the direct methods, Topic Unit Methods will be activated, presenting information on groundwater tracers (and related sampling methods). Otherwise, the user should proceed with estimation of hydraulic conductivity.

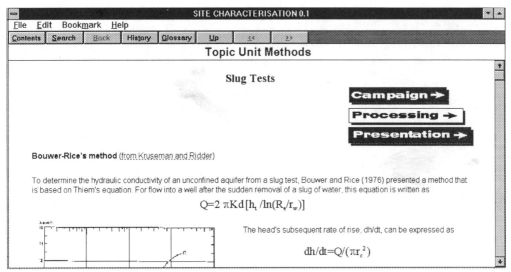

Figure 4.21   Topic Unit <u>Methods</u>, example b)

Once the Guide is consulted, the method can be chosen. By clicking the arrow next to the selected method, a new (sub)topic will be open showing specifics of the method (Figure 4.21).[17] After mastering the topic, the user can choose between 'Campaign', 'Processing' and 'Presentation'. If any conductivity data exist, a decision on additional site investigations (Campaign) is usually taken after Processing and Presentation. If field data are already processed (i.e. conductivity estimated) Presentation will take place.[18] In REGIS, hydraulic conductivity data are stored (and can be presented) in a form of so-called litho-k-value columns. These columns form the basis for interpolation and 'translation' of lithostratigraphic into hydrogeological information (Section 4.3). Once interpolation and translation are completed, i.e. once a conceptual model of the GWS is developed, the model can be stored and presented in REGIS (the so-called Geohydrological Layer Model).

Desk-top Processing is the estimation (calculation) of hydraulic conductivity from field data. (In the example followed here, field data are obtained by applying the slug test.) By clicking the button 'Processing', software 'Single Well Solution Software' (StreamLine Groundwater Applications, 1996) will be activated. One of the software output screens is shown in Figure 4.22.

---

[17] In the case of 'bail and slug tests', only one subtopic is created in the prototype (Figure 4.21), covering the Bouwer-Rice method; a similar subtopic can be developed for the Cooper-Bredehoeft-Papadopulos method, the Hvoslev method, etc.

[18] In this case, the user will neither carry out the Slug Test (Campaign), nor process the field data; the only task left is to present the conductivity values. Nevertheless, it is very important that user goes through the Topic Unit <u>Method</u> (next to <u>Objectives</u> and <u>Parameters</u>), and learns (is reminded of) the specifics of the method used for a parameter estimation.

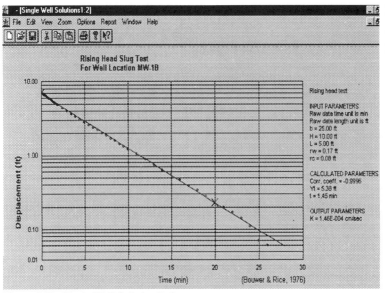

Figure 4.22   Single Well Solution Software, an output screens

Finally, the SCM can assist the user in planning (additional) site investigations. First of all, the user is 'equipped' with knowledge on parameter(s) to be defined and method(s) to be applied. In addition to that, a sheet of the electronic notebook can be prepared for each parameter, as described in Section 4.3 (Task Unit CAMPAIGN). If, for example, a slug test is planned, then the sheet should contain a form on which the field data (head, well radius, screen length, etc.) could be recorded. The form can be accompanied by some simple graphics where field data to be measured (Figure 4.23) or equipment to be used (Figure 4.20) are sketched. (Naturally, the content of the notebook sheet is also defined by the method used). Most of the sheets should contain a local scale map with locations of already conducted and planned

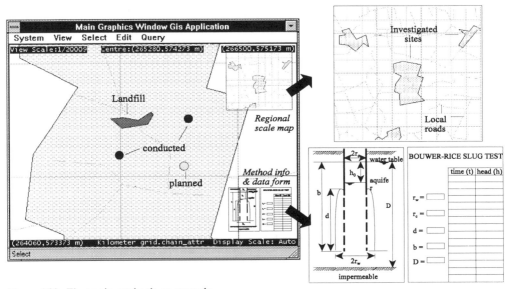

Figure 4.23   Electronic notebook, an example

investigations; an (orientational) regional scale map could be included as well (Figure 4.23). The notebook prints should be taken along to the investigated sites.

The example presented in this section does not exhaust information on the hydraulic conductivity that can be included in (and afterwards supported by) the SCM. The purpose of the example given is to initiate comments and ideas that would improve and enlarge the SCM.

4.5.2 Solubility

Information on chemical solubility is required at different stages of evaluation of a groundwater pollution situation. At the very beginning of the characterisation, solubility can be used as a first (pessimistic) estimate of concentration. Besides, it can also be used for derivation of another parameter: the organic carbon/water partition coefficient (K_{oc}).[19] Information on solubility contained in the Module and the way in which the encapsulated knowledge can be used are described below.

In the preliminary stage of the Site Characterisation, when no groundwater sampling has been conducted, very little can be said on pollutants that (might be) spread within the site. Nevertheless, the user of the SCM can consult the topic 'Sources of groundwater pollution' (see Objectives, Section 4.3) and find out which contaminants are usually associated with the (potential) pollution source located at the site. Already at that stage, mobility, toxicity and solubility should be checked in order to select a preliminary indicator parameter for an initial estimation of the pollution situation. (This estimation is required for prioritisation of the sites for further investigations.) Since no data on concentration are available, solubility of the chemical can be used as a pessimistic upper limit of concentration.

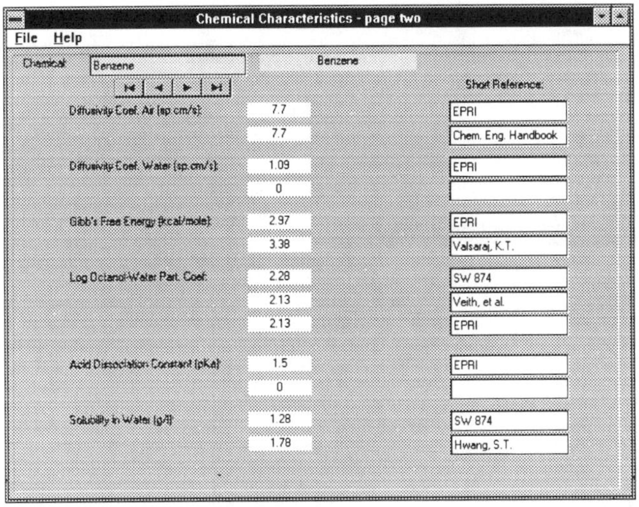

Figure 4.24 Chemical DataBase (CDB)

The topic 'Solubility' (Topic Unit <u>Parameters</u>) contains only a few sentences on the role of solubility in evaluation of groundwater pollution. This topic, however, contains links to: Topic

---

[19] K_{oc} is needed for estimation of retardation due to sorption; the retardation factor is used by the Vulnerability Assessment Module (VAM), as well as by the module for Groundwater Pollution Modelling (GMM).

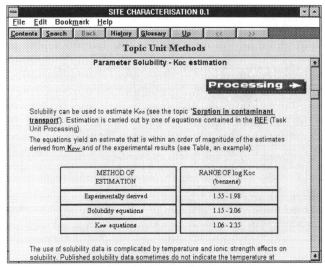

Figure 4.25   Topic Unit Methods, example c)

Unit <u>Processes</u>, the Chemical DataBase (CDB) and Topic Unit <u>Methods</u>. By clicking on highlighted text 'processes' the topic 'Solubility' (Topic Unit <u>Processes</u>) will be shown on the screen.[20] In this topic, definition of solubility is given, together with its importance, ranges, units, and the factors that influence solubility (temperature, salinity, pH, etc.). Once familiarised with the content of this topic, the user should go back to Topic Unit <u>Parameters</u>, and choose one of two buttons: 'CDB' or 'Methods'.

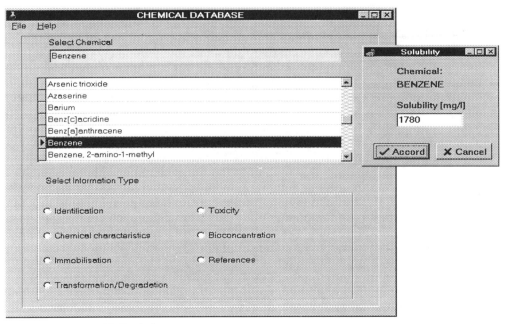

Figure 4.26   CDB and RE1 windows

If a solubility value is needed as a first (worst-case) estimation of the concentration, then the Chemical DataBase should be activated (the button 'CDB'). A window of the CDB that contains

_____

[20] In the Topic Unit <u>Processes</u>, the topic 'Solubility' is a (sub)topic of 'Factors influencing sorption', and this topic is a (sub)topic of 'Sorption' (see Figure 4.7).

one or more solubility value(s) will appear on the screen ('Chemical characteristics - page two', Figure 4.24).

The button 'Methods' activates the topic in which the method for estimation of K_{oc} from solubility values is described (Figure 4.25). The user is again advised to consult the Topic Unit Processes, where extensive information on K_{oc} and its role in estimation of retardation can be found. By clicking the button 'Processing', the CDB will be activated (the main window will appear), but this time together with an adjacent window, containing the solubility value (Figure 4.26). The adjacent window belongs to another (simultaneously activated) software application called RE1.[21] At this point the user can accept the displayed value or type-in an alternative. Once the value is accepted (button 'Accord'), a new RE1 window will be displayed (Figure 4.27). At the left side of the window several empirical equations are given (by author name) which can be used for K_{oc} estimation from solubility.[22] Calculated (log) K_{oc} values appear at the right side of the window. If the user selects more then one equation, basic statistics of estimated K_{oc} values will be calculated as well; the results of the statistical analysis can be useful when selecting one of estimated values.[23]

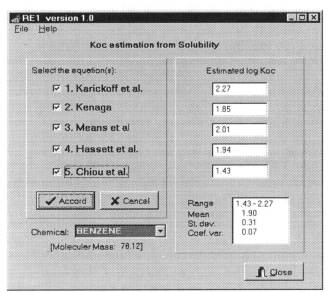

Figure 4.27 K_{oc} estimation from Solubility

---

[21] RE1 is software that contains a procedure for estimation of the retardation factor (see Section 6.5.2). It was developed for an MS Windows (NT 3.51) environment by using object-oriented Borland Delphi Developer, a tool based on Object Pascal as a programming language. Dynamic Data Exchange (DDE) is established between the RE1 and CDB for selection of the solubility value.

[22] The window also displays a molecular mass of the chemical that is automatically retrieved from the CDB. The molecular mass is required for conversion from milligrams to moles or to mole fraction, units that are used in some of the included equations.

[23] Information on chemicals that are used for derivation of equations should be found and placed next to the equations (on the window shown in Figure 4.27). Then the user should select an equation that is derived on the basis of a chemical similar to the one under study. This would certainly increase reliability of the estimated K_{oc}. (Information on chemicals used are already included in the procedure for determination of K_{oc} from K_{ow} - see Section 6.5.2).

K_{oc} is one of the parameters related to the objective 'evaluation of groundwater pollution' (Figure 4.6). If the K_{oc} parameter is approached 'directly' (i.e. not via 'Solubility'), the user could choose between estimation of K_{oc} from K_{ow} or from the Solubility.[24] If the latter is chosen, the topic 'Solubility-K_{oc} estimation' will be displayed on the screen (Figure 4.25).

Once K_{oc} is estimated, the user should proceed with estimation of the retardation factor. The latter does not have to be estimated during the Site Characterisation, because it is not required for the characterisation, but for the vulnerability assessment (the VAM module) and pollution modelling (the GMM module). However, when the estimation of the retardation factor is needed, the SCM will be activated (from the VAM or from the GMM), because this module contains knowledge on retardation (and sorption, solubility, K_{oc}, porosity, etc). Therefore, if the assessment is to be carried out after the characterisation, the retardation factor can already be estimated during the Site Characterisation.

Annex: Glossary

Analysis - is a desk-top step of SC and a Task Unit in the SCM. SC begins with the analysis of a current groundwater pollution problem; the user has to set up objectives, to define parameters required for fulfilment of objectives and to choose methods that are (or could be) applied in order to determine parameters. Task Unit ANALYSIS consists of four topic units: Objectives, Parameters, Methods and Processes.[25]

The GroundWater System (GWS) - is a hydrogeological environment that contains a part of an aquifer or the aquifer in its full extent. A distinction is made between GWS and aquifer because the GWS is defined by the scale of a groundwater pollution problem that does not have to resemble natural aquifer borders. (The GWS is an environment where groundwater pollution takes place.) If a GWS contains more (or parts of) aquifers separated with aquitards, it should be referred to as the complex GWS. To define the GWS means definition of the model components: system matrix, system boundaries and boundary conditions and internal conditions.

Interpretation - is a desk-top step of SC and a Task Unit in the SCM. It denotes synthesis and explanation of information obtained by analysis (and processing).

---

[24] If the former (i.e. K_{ow}) is chosen the adjacent window (Figure 4.26) will contain K_{ow} value from the CDB. Subsequently, the window 'Koc estimation from Kow' (Figure 6.14), similar to one shown in Figure 4.27, will be displayed.

[25] In practice, the term 'analysis' is often dropped out and the term 'processing' combines 'logical processing' (i.e. analysis) and physical (numerical) processing. In the SCM, Analysis and Processing are two distinct tasks and, therefore, two distinct Task Units. (Analysis is usually followed by Synthesis; in the Module, synthesis is a part of the INTERPRETATION Task Unit.)

Presentation - is a desk-top step of SC and a Task Unit in the SCM. It denotes implementation of procedures for graphical (or tabular) presentation of information that is obtained directly from the field or as a result of processing.

Processing - is a desk-top step of SC and a Task Unit in the SCM. It denotes implementation of procedures for processing of quantitative (numerical) information.

Site Characterisation (SC) - is the first task to be carried out while dealing with point-source groundwater problems. It includes gathering, (analysis and) processing and presentation (and interpretation) of information on a polluted site. The desk-top part of SC encompasses the preparation of field and laboratory activities, and analysis, processing, presentation and interpretation of results obtained. The Site Characterisation Module (SCM) contains qualitative information (knowledge) and links with processing tools, and is developed to assist a user in carrying out the desk-top part of SC.

Site Characterisation Module (SCM) - is a hypertext-based software tool developed to assist a user in carrying out the desk-top part of SC (i.e. a preparation of field and laboratory activities, and analysis, processing, presentation and interpretation of results obtained).

System Matrix - refers to the lithological and structural characteristics of the geological environment. Layers of sedimentary rocks are called strata, so the geological units within the matrix are often referred as lithostratigraphic units. The units are determined by regionalisation of field data and, subsequently, translated into hydrogeological units.

System Boundaries and Boundary Conditions - system boundaries are the spatial boundaries of a GWS. (The boundaries can be physical, e.g. impermeable layer or surface water; or hydraulic, e.g. groundwater divide and streamlines.) System boundary conditions refer to the flow and transport conditions at the boundaries of a GWS.

System Internal Conditions - refer to the flow and transport conditions within the GWS. They are determined (exclusively) by the characteristics of the GWS (i.e. characteristics of pollutants play no role) and represented by the set of parameters.

Task Unit - is an organisational unit of the SCM. Task units consist of a single topic unit and commands merged into the topic unit (the exception is Task unit ANALYSIS that includes four topic units and no commands). Each task unit is dedicated to one of the main task of the Site Characterisation.

Topic Unit - is an organisational unit of the SCM. Topic units are composed of sets of topics. Each topic unit is dedicated to one of the main topics of the Site Characterisation.

Topics - are hypertext-based knowledge representation forms that contain information presented in textual and graphical forms (images, diagrams and tables) and links with other related topics and other software. units contain text, images, graphics and tables.

5. KNOWLEDGE-BASED MODULE FOR VULNERABILITY ASSESSMENT

5.1 Introduction

In general terms, groundwater Vulnerability Assessment (VA) can be described as a procedure for the quick assessment of a groundwater pollution potential. It is based on intrinsic aquifer characteristics, though contaminant characteristics and management practice can also be taken into account.

In the context of investigations of groundwater pollution potential (at a local scale), VA is seen as a step that follows characterisation of the investigated site. Accordingly, VA uses the results of the site characterisation, yielding a first assessment of the pollution potential. The results of VA provide a basis for further investigations and/or assessment, and a means for comparison of pollution potentials.

VA methodologies developed for point source pollution problems are not sufficiently applied in practice. One of the reasons is certainly a lack of consensus over factors (parameters) that should be included and their relative importance. Additionally, some of the obstacles are related to very few methodologies being electronically encapsulated and integrated with other software. Nevertheless, contemporary information technology provides the means for integration of the tools and adaptive electronic encapsulation of the knowledge that are required to assess groundwater vulnerability to pollution.

The research project 'Decision Support System for Groundwater Pollution Assessment' included development of knowledge-based software for VA. According to the notion of integrated information, the Vulnerability Assessment Module (VAM) had to be designed and developed in such a way as to make full use of the other software integrated into the DSS. Therefore, a number of VA methodologies for local-scale pollution problems were critically reviewed in order to select the most suitable for integration into the DSS. This review, a summary of which is given in Section 5.2 of this chapter, has showed that none of investigated methodologies fulfilled all or even most of the posed requirements. A new methodology, that will be described in Section 5.3, was developed and subsequently encapsulated into the VAM and integrated into the DSS (Section 5.4). The VAM was tested on a case-study, discussion of which in Section 5.5 will illustrate the advantages of integrated information.

5.2 Groundwater vulnerability in theory and practice

The concept of groundwater vulnerability was outlined in Section 2.3.3, acknowledging the difficulty to define precisely 'vulnerability assessment'.[1] The contemporary use of the term 'Vulnerability Assessment' was intensively discussed during the VAM design and development. It appeared that a lack of sharp differentiation between vulnerability assessment and risk analysis causes most of the terminological inconsistency.[2] Moreover, the concept of vulnerability assessment has been extended, so that intrinsic, as well as specific (or integrated), vulnerability can be distinguished. The terms 'Vulnerability Assessment' and 'Vulnerability Mapping' are not synonymous; the mapping is (just) one of the VA approaches. Still, 'vulnerability means different things to different people', as stated by National Research Council (1993).

Several VA methods developed for assessment at a local scale were selected for potential integration into the DSS for Groundwater Pollution Assessment. Since the spatial variability of some parameters should not be completely disregarded even at local scale, one of the regional assessment methods (DRASTIC) was also included in the selection. Selected methods were described, or at least mentioned in Sections 2.2 and/or 2.3.3. Some additional information, relevant for the coming comparison, is presented below.

The *LeGrand System* (Legrand, 1980) is a numerical rating system developed for evaluation of the potential for groundwater contamination from waste disposal sites. The system focuses on weighting four key parameters that describe site hydrogeology: distance to a water supply, depth to a water table, hydraulic gradient and permeability-sorption. The parameter ratings are gathered independently and the probability of contamination is determined by correlating a site description with a synthetic standard.

The US EPA (1992) developed the *Hazard Ranking System* (HRS) to be used in the assessment

---

[1] Some of the suggested definitions are listed below:

 – 'the tendency or likelihood for contaminants to reach a specified position in the groundwater system after introduction at some location above the uppermost aquifer', U.S. National Research Council, The Committee on Techniques for Assessing Groundwater Vulnerability (1993);
 – 'the ability of the system to cope with external impacts, both natural and anthropogenic, which affect its state and character in time and space', Sotornikova and Vrba (1987);
 – 'the sensitivity of its quality to anthropogenic activities which may prove detrimental to the present and/or intended usage-value of the resource', Bachmat and Collin (1987);
 – 'an intrinsic property of a groundwater system that depends on the sensitivity of that system to human and / or natural impacts', Vrba and Zaporozec (1994);
 – 'the risk of chemical substances - used or disposed on or near the ground surface - to influence groundwater quality', Villumsen et al (1982).

Finally, Adams and Foster (1992) defined the vulnerability of an aquifer to contamination as being a function of: 'the inaccessibility of the saturated zone, in a hydraulic sense, to the penetration of contaminants', and 'the attenuation capacity of the strata overlying the saturated zone as a result of physicochemical retention or reaction of contaminants'.

[2] In the classification given by the Vrba and Zaporozec (1994) the 'Standardised system for evaluating waste disposal sites' (LeGrand, 1980) represents a whole group of vulnerability assessment methods. Authors of the 'Hazard Ranking System' (USEPA, 1992) consider the System also as a vulnerability assessment method.

of relative threat associated with the release, or potential release, of hazardous substances from a waste site. The HRS incorporates a considerable number of parameters that are, however, divided among four possible pollution pathways: ground water, surface water, soil and air, hence HRS is considered more a screening rather than an assessment tool. Each pathway is determined by three groups of factors: likelihood of release, waste characteristics and targets. The HRS is encoded in PRESCORE software developed for an MS DOS environment and integrated with a chemical database.

GEOTOX (Wilson et al, 1987; Mikroudis, 1988) is software developed for hazardous site evaluation. The encapsulated methodology considers four groups of factors: contaminant or waste characteristics, waste management practice, pathways and targets or receptors. The parameters in GEOTOX are summed (aggregated) in such a way as to include confidence measures. GEOTOX is encoded (using Prolog programming language) as a rule-based system for an MS DOS environment. The software is transparent, i.e. the basic structure and majority of the rules can be reviewed by the user. Moreover, the user is able to change parameter ratings, confidence in parameters and to introduce new rules.

DRASTIC (Aller et al, 1987) is a standardised system for evaluating groundwater pollution potential using mappable units of seven major hydrogeological factors that affect and control groundwater movement: depth to water, net recharge, aquifer media, soil media, topography, impact of the vadose zone media and hydraulic conductivity of the aquifer. Relative ratings and weights are introduced to combine the factors into a numerical value called the DRASTIC index. The result of the assessment is a vulnerability map. DRASTIC is extensively used not only to assess a threat from diffuse pollution sources, but also in the planning of groundwater monitoring programmes and selection of waste disposal sites. Development of GISs (and coupling with them) have made the use of this methodology especially attractive.

The examined vulnerability methods developed for local scale problems were fairly comparable with respect to the factors they include; in one way or another, all of them consider the characteristics of waste, hydrogeological environment and potential target. Besides, all these methods result in a score that is meant for comparison with other investigated sites (i.e. 'external' comparison). A DRASTIC outcome, however, provides a means for external, as well as 'internal' comparison (a DRASTIC index score and vulnerability map, respectively).

The vulnerability methods were examined with respect to suitability for integration into the DSS. The results, summarised in Table 5.1, show that none of the examined methodology fulfilled all the requirements. Therefore, a decision was made to develop a new methodology that would combine the best features of existing methodologies (regarding design, selection of factors, relations among factors, etc.), fulfil posed requirements and become fully integrated into the DSS.

Table 5.1 Comparison of vulnerability assessment methodologies

Requirements relevant for integration into the DSS	Methodology			
	LeGrand	**HRS**	**GEOTOX**	**DRASTIC**
Developed for a local scale pollution problems	✔	✔	✔	-
Developed for assessment (i.e. not for screening)	✔	-	✔	✔
Considers spatial variability of parameters	-	-	-	✔
Electronically encapsulated (i.e. software developed)	-	✔	✔	✔
Contains knowledge (if electronically encapsulated)	-	-	✔	-
Provides evidence of previous integration	-	✔	-	✔

5.3 Description of methodology

The information acquired from the literature and software analyses was used in the selection of factors that define the vulnerability of the aquifer, as well as vulnerability of a specified target within the aquifer. The main criteria for the selection were factor importance and availability of information on the factor. Additionally, the methodology would be oversimplified if only a few factors were selected, whereas the involvement of all the encountered factors would cause difficulties in terms of complexity (factor comparison and grouping) and information availability. Several factors that form a core of the methodology were chosen to be compulsory for the assessment, while all the others can be neglected, if the information on the factor is not available or if the user finds the factor insignificant. The user can add a new factor, and all the factors can be modified (see Section 5.4).

The selected factors were classified into four groups:

1. Likelihood of release and aquifer pollution (LI factor);
2. Contaminant severity (CS factor);
3. Pathway activity (PA factor), and
4. Target exposure (TE factor).

(For the sake of terminological clarity, the groups are denoted as 'factors', while the individual items within the groups are called 'parameters'.) In order to provide a unique comparison among the investigated sites, the factors are eventually joined into a unique OAS - Overall Assessment Score (see Figure 5.1).

Each parameter has at least two attributes: 'value' and 'rating'. Some parameters also have 'confidence' as a third parameter.. The attribute 'value' contains real values of a parameter, given in a descriptive form or as a numerical range (see Table 5.2 for an example).

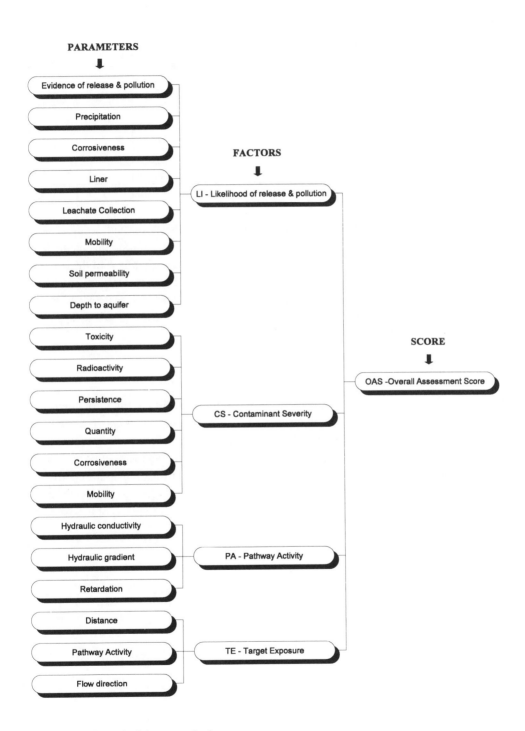

Figure 5.1 The VA methodology organisation

Table 5.2 Examples of parameter attributes

LI factor parameter: LINER			LI factor parameter: PRECIPITATION		
values	rating	confidence	values [mm/y]	rating	confidence
composite	1	0.09	<25	1	0.18
clay or other liner persistent to organic compounds	3	0.09	25 - 150	3	0.18
geomembrane (synthetic or concrete liner)	5	0.09	150 - 400	5	0.18
asphalt based liner	7	0.09	400 - 800	7	0.18
no liner used	9	0.09	> 800	9	0.18

The attribute 'rating' was introduced in order to allow comparison and grouping of the parameters. The 'confidence' is virtually a probability of occurrence, but in this methodology, it is considered primarily as a measure of importance (rather then reliability) of the parameter. Detailed description of the attributes and the way that attributes (parameters) are grouped will be given below for each factor separately. Some issues related to the newly introduced methodology are then discussed. Section 5.3 is rounded off with the testing of the 'aggregation procedure', which is the core of new VA methodology.

5.3.1 Likelihood of release and aquifer pollution (LI factor)

Procedure This factor is dedicated to the (potential) source of pollution. Eight parameters were selected for the assessment of likelihood of release and aquifer pollution (Table 5.3). The parameters have four or five different values, with the exception of the parameter 'Evidence..' where one of the six different values can be chosen. The rating and the confidence attributes were assigned to each parameter. The LI factor was determined by adding up the contributions of each individual parameter. The contribution of each parameter is defined by selected a rating (R_i) and standard deviation (S_i), so that the combined (aggregated) rating (R) of two parameters was defined as:

$$R = \frac{R_1 * S^2}{S_1^2} + \frac{R_2 * S^2}{S_2^2} \tag{5.1}$$

where

$$S = \frac{(S_1 * S_2)}{SS} \tag{5.2}$$

and

$$SS = \sqrt{(S_1^2 + S_2^2)} \tag{5.3}$$

Table 5.3 Likelihood of release and aquifer pollution (LI factor) parameters

LI factor parameter: SOIL PERMEABILITY		
values	rating	confidence
impermeable (clay >50%)	1	0.18
relatively impermeable (30% <clay <50%	3	0.18
relatively permeable (15% <clay <30%)	6	0.18
very permeable (clay <15%)	9	0.18

LI factor parameter: MOBILITY		
values	rating	confidence
low	2	0.36
moderate	4	0.36
high	6	0.36
very high	8	0.36

LI factor parameter: EVIDENCE OF RELEASE AND AQUIFER POLLUTION		
values	rating	confidence
no evidence of release	1	0.36
No evidence - conviction	3	0.36
indirect evidence	5	0.50
positive proof from direct observation	7	0.50
positive proof from laboratory analyses	9	0.50
evidence of aquifer pollution	☆	☆

LI factor parameter: DEPTH TO AQUIFER		
values [m]	rating	confidence
< 1	1	0.18
1 - 5	3	0.18
5 - 20	5	0.18
20 - 50	7	0.18
> 50	9	0.18

LI factor parameter: LINER		
values	rating	confidence
composite	1	0.09
clay or other liner persistent to organic compounds	3	0.09
geomembrane (synthetic or concrete liner)	5	0.09
asphalt based liner	7	0.09
no liner used	9	0.09

LI factor parameter: PRECIPITATION		
values [mm/y]	rating	confidence
<25	1	0.18
25 - 150	3	0.18
150 - 400	5	0.18
400 - 800	7	0.18
> 800	9	0.18

LI factor parameter: LEACHATE COLLECTION		
values	rating	confidence
adequate collection and treatment	2	0.09
inadequate collection or treatment	4	0.09
inadequate collection and treatment	6	0.09
no collection or treatment	8	0.09

LI factor parameter: CORROSIVENESS		
values	rating	confidence
6 < pH < 9	2	0.09
5 < pH < 6 or 9 <pH <10	4	0.09
3 < pH < 5 or 10 <pH <12	6	0.09
1 < pH < 3 or 12 <pH <14	8	0.09

In the next step, the aggregated rating (R) from the previous step was regarded as R_1 in the equation (1), while a new parameter contribution (to be added) was assigned to the R_2. The standard deviation was obtained from the confidence level, i.e. probability of occurrence, assuming a normal distribution of the parameter (see Section 5.3.7).

The outcome of the assessment (i.e. selection of values and aggregation) is the LI factor score, that can take any value between minimum and maximum parameter ratings (e.g. from 1 to 9). The score was eventually divided into four ranges, providing a descriptive rating of assessed likelihood of release and aquifer pollution (Table 5.4). The confidence threshold was also introduced, requiring a minimum number of parameters to be introduced into the procedure in order to secure reliability of the assessment. The minimum number of parameters required was defined by their importance.

Table 5.4 Examples of factor rating

| LI FACTOR SCORE and RATING ||
score	rating
< 3	low
3-5	moderate
5-7	high
> 7	very high

The outlined procedure was extensively tested with respect to allocated ratings, confidence, threshold confidence and the LI factor score (Section 5.3.7).

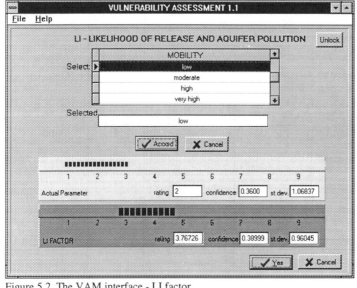

Figure 5.2 The VAM interface - LI factor.

User Interface The user interface for the LI factor is shown in Figure 5.2. By clicking with the mouse in the upper box at one of the offered parameter values, the value will appear in the lower box, and its rating, confidence and standard deviation will be shown in the upper scale panel. The rating will be positioned at the middle of the horizontal bar whose length indicates the confidence expressed by the standard deviation. In this way, the transparency of the software is assured, providing the user with an overview of all the attributes involved.

After the value is selected (button 'Accord' clicked), the aggregation will take place according to the above-mentioned procedure. Aggregated rating, confidence and standard deviation will be shown at the lower scale panel. Simultaneously, a new parameter will appear in the upper box. The process will continue until all the parameter ratings are summed up. Then the boxes and the upper scale panel will disappear from the screen and the assessed descriptive rating of the LI factor will appear (e.g.'Likelihood of release and aquifer pollution is moderate').

There are two exceptions related to the parameter 'Evidence of release and aquifer pollution'. If the user confirms that evidence of release exists, the parameters 'Liner' and 'Leachate collection' will not be taken in consideration. If the user confirms aquifer pollution, the assessment of the LI Factor will be terminated. The highest rating of the LI factor (very high) will then be used in the calculation of the OAS.

5.3.2 Contaminant severity (SC factor)

The same procedure and the same design of the User Interface were used for the assessment of contaminant severity. The parameters involved in assessment and their attributes are shown in Table 5.5.

Table 5.5 Contaminant Severity (CS factor) parameters

SC factor parameter: TOXICITY			SC factor parameter: RADIOACTIVITY		
values	rating	confidence	values	rating	confidence
SAX level 0 or NFPA level 0	2	0.21	At or below background level	2	0.21
SAX level 1 or NFPA level 1	4	0.21	1 to 3 times background level	4	0.21
SAX level 2 or NFPA level 2	6	0.21	3 TO 5 times background level	6	0.21
SAX level 3 or NFPA level 3 or 4	8	0.21	Over 5 times background level	8	0.21
SC factor parameter: PERSISTENCE			Li factor parameter: MOBILITY		
values	rating	confidence	values	rating	confidence
Easily biodegradable compound	2	0.15	low	2	0.36
Straight chain hydrocarbons	4	0.15	moderate	4	0.36
Substituted and other ring compound	6	0.15	high	6	0.36
Metals, polycyc. comp. and halog. hydrocarbon	8	0.15	very high	8	0.36

Table 5.5 continued

SC factor parameter: QUANTITY			SC factor parameter: CORROSIVENESS		
values	rating	confidence	values	rating	confidence
Less than 250 tons	2	0.21	$6 < pH < 9$	2	0.09
251 to 1000 tons	4	0.21	$5 < pH < 6$ or $9 < pH < 10$	4	0.09
1001 to 2000 tons	6	0.21	$3 < pH < 5$ or $10 < pH < 12$	6	0.09
Greater than 2000 tons	8	0.21	$1 < pH < 3$ or $12 < pH < 14$	8	0.09

Two parameters, namely 'Mobility' and 'Corrosiveness' are seen to be involved in assessment of both the LI and SC factors. Originally, the idea was to use these two factors for assessment of the SC factor only, if assessment of LI factor was skipped by the user. Nevertheless, it was decided to treat the LI and SC factors as completely independent vulnerability indicators.

5.3.3 Pathway activity (PA factor)

Procedure The PA factor assesses a readiness of a hydrogeological environment to transport a pollutant. If the environment is more 'active', the pollutant will be easily transported. The investigated area is discretised grid-wise, and the PA factor is calculated for each cell of the grid[3]. In the vertical direction, a cell extends throughout the relevant aquifer(s). Possible variation in parameter values within the cell (in all directions) are averaged. If the assessment is carried out for a leaky aquifer, then the PA factors are defined separately for the aquitard and the leaky aquifer and subsequently combined.

The pathway activity factor is composed of three parameters that were found to be the most relevant for the assessment, namely: hydraulic conductivity (k), gradient (I) and retardation (R). The assessment of the PA factor is actually an estimation of transport velocity for each cell of the investigated area:[4]

---

[3] The investigated area is defined during the Site Characterisation. It defines the dimensions of the investigated groundwater system in a horizontal direction. A vertical extension of the groundwater system is defined by a depth of the aquifer(s) that comprise the groundwater system.

[4] Darcy's law expresses the rate of flow through a porous media as:
$$Q = k * I * A$$
where 'Q' is flow rate and 'A' is cross-sectional area of flow. The law can be rewritten as:
$$Q/A = v = k * I$$
where v = specific discharge or Darcy velocity. Since groundwater flows only through the voids, it must follow flow paths around the solids. The actual or 'seepage' velocity is given as:
$$v_s = v / n_e$$
where 'v_s' is seepage velocity, 'n_e' is effective porosity and 'v' is specific discharge. For pollutant transport, the
(continued...)

$$PA = \frac{k*I}{R} \tag{5.4}$$

Hydraulic conductivity was assigned to each cell by extrapolating information obtained from the site. A cell gradient was defined as the largest gradient that occurs between the actual cell and (usually eight) neighbouring cells. Retardation (R) was calculated as

$$R = 1.0 + \frac{P_b * K_d}{n} \tag{5.5}$$

where 'Pb' is the dry bulk density of the medium and 'K_d' is a partition coefficient between sorbed and unsorbed part of the pollutant[5]. The conductivity and gradient are solely intrinsic aquifer characteristics, while the retardation is defined by the characteristics of the aquifer, as well as by the pollutant characteristics.

Two kinds of rating were introduced for each of the parameters: internal and external. The internal rating was obtained by dividing the actual range of a parameter (log values) into ten classes and assigning the parameter values to the classes[6]. Absolute range (the most extreme observed or calculated ever) of parameters was used to perform the external rating. External rating is required for comparison of the polluted sites. The PA factor as a whole (obtained by applying Equation 5.4) is also rated internally.[7]

Three grid maps are produced for each of the parameters showing actual values, internal rating and external ratings. An actual value map and an internal rating map are also made for the PA factor. The latter maps are the main result of the PA factor assessment, and the actual value map

---

[4](...continued)
velocity is expressed as:
$$v_p = v_s / R$$
where 'R' is the retardation. Combining previous equations, pollutant transport velocity is given as:
$$v_p = k*I / (n_e * R)$$
The porosity term n, mentioned in the above relation is already included in the expression for retardation and could be neglected in order to define the PA factor.

[5] K_d is further expressed as
$$K_d = K_{oc} * f_{oc}$$
where 'K_{oc}' is an organic carbon/water partition coefficient and 'f_{oc}' is an organic carbon content. K_{oc} is usually derived from K_{ow} (octanol/water partition coefficient) or from the solubility values.

[6] Actual range is range of values observed at an investigated site or calculated using the site information.

[7] Equation 5.4 is, however, not used for the external PA factor rating, because the range created by introducing extremes for K, I and R into the equation is so wide, unreliable and practically worthless (the chance of all the values having the same rating is very high). Therefore, the aggregation procedure is applied to the external PA factor rating.

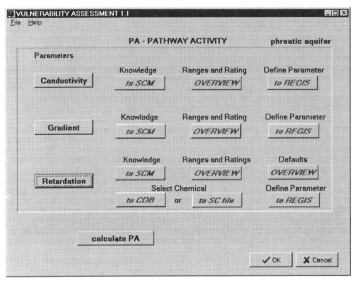

Figure 5.3 The VAM interface - PA factor

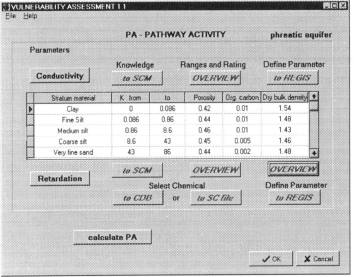

Figure 5.4 The VAM interface - PA factor defaults

is used for determination of the Potential Hazard Map (see Target Exposure factor). External parameter rating maps are needed for determination of TE factor (external comparison), as well as Overall Assessment Score (OAS).

User Interface After the decision is made as to whether a phreatic or leaky aquifer is to be assessed, the user is prompted to the window shown in Figure 5.3. By pressing the buttons 'Conductivity', 'Gradient' and 'Retardation', the buttons related to these parameters will appear on the screen.[8] There are at least three buttons related to each parameter: 'to SCM', 'Overview' and 'to REGIS'. The button 'to SCM' establishes the connection with the Site Characterisation Module (SCM) where the qualitative information about the actual parameter is stored (see, for example, Figure 4.18). The button 'Overview' offers an overview of ranges and ratings used for the external ratings. An additional button 'Overview' related to the parameter retardation prompts the user to the defaults used for calculation of this parameter

---

[8] Calculation of the retardation requires K_{oc} that can be obtained from the Site Characterisation result file ('to SC file' button) if the related chemical is selected as an indicator parameter during the site characterisation. The other option is to activate the RE1 module which will perform the calculation (see Section 6.5.2).

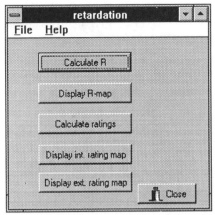

Figure 5.5 VAM 'execution' window, an example

(Figure 5.4). If the button 'to REGIS' is clicked, a new, smaller windows will appear on top of the main one (Figure 5.6 is an example). Via these windows (three in total) the calculation and display of the maps is carried out. The maps will be shown in Section 5.5 (Case study).

5.3.4 Target exposure (TE factor)

Procedure The TE factor is defined as an aggregated (gross) pathway activity between a potential source and a potential target. It is used for internal (within the site), as well as external comparison.

The TE factor is the main outcome of the 'internal' assessment, being the ratio of PA and distance between the specified source and a number of potential targets.[9] All the grid cells with the negative (downwards) gradient (in respect to the source) are considered as potential targets. The outcome is presented in a form of the map called the 'Potential Hazard Map'. (The map is made under the assumption that groundwater takes the shortest distance while flowing between the source and the target; however, this does not always have to be the case.) Two additional maps have been prepared in order to enhance the assessment of vulnerability of the potential target(s): a Groundwater Pathway Map and a Distance Map. In the Groundwater Pathway Map a pathway is shown that starts at the location of the source and then follows the steepest gradient from the source. If the information on gradients is correct, the pathway should be a fairly reliable representation of groundwater flow direction. The Distance Map shows distances between the source and the targets; beside actual values, internal and external ratings are calculated as well.

The three maps described can be used in order to estimate exposure (or assess vulnerability) of a target to the pollution at any point within the investigated site. The location (coordinates) of the potential source can be changed and the assessment repeated. This makes a VAM also a convenient tool for planning of, for example, new disposal sites (potential source) or new well fields (potential target).

The *User Interface* for the TE factor is very simple. Only one screen (Figure 5.6) is used as a joined interface for determination of both TE factor and Overall Assessment Score (OAS). One

---

[9] TE was obtained by summing up the ratios of the PA_i and distance (d_i) for each cell 'i' that is included in a linear pathway between the source and the target. TE is inverse proportional to the travel time (t):

$$t = \sum_{i=1}^{i=n} \frac{d_i}{v_p}$$

where v_p is transport velocity.

of the reason for this combination was use of an external rated TE factor for determination of the OAS. In order to complete internal assessment, the user has only to select coordinates of the potential source and subsequently to analyse the maps produced. Determination of the OAS (OAS is used for external comparison) is discussed below.

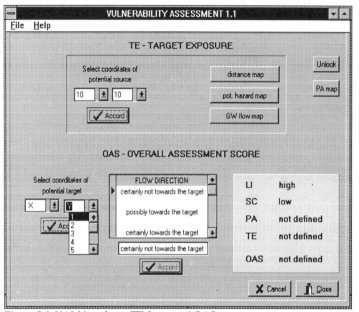

Figure 5.6 VAM interface - TE factor and OAS

5.3.5 Overall assessment score (OAS)

Procedure   The OAS combines all four factors in order to provide comparison between sites. The LI and CS factors can be used straightforwardly for the comparison, but a comparison of TE for all the possible targets would be rather meaningless. Therefore, one particular target can be selected (for instance, the existing target that has the highest exposure according to the internal rating) and its exposure estimated. The external TE factor rating is a combination of the PE factor, distance (both external ratings) and groundwater flow direction (obtained from the GPM). Eventually the OAS is defined as a sum of the factors

$$OAS = LI + SC + PA + TE \qquad (5.6)$$

by using the aggregation procedure.

User Interface   The user has to define the coordinates of the target and the general direction of the groundwater flow estimated from Groundwater Pathway Map. This step completes the assessment.

5.3.6 Discussion

The following issues can be distinguished as the most important for setting up a rating-based methodology: definition of parameters values, application of the ratings and definition of relations between the parameters.

The parameter values can easily be defined for quantitative parameters with known extremes. Under the assumption that their importance (or impact) changes (increases or decreases) linearly between the extremes, the ratings can also be easily assigned (e.g. from 1 to 10). However, not all the quantitative parameters have clearly defined extremes; what should be, for instance, maximal distance at which a possible threat of pollution should still be taken into account? This issue was one of the reason for the introduction of internal ratings for the pathway activity parameters (where a range of a parameter values is defined by the extremes obtained from the actual site). Besides, the absolute range can be so wide that all the parameter values observed at the site can easily attract the same rating.

Not all the descriptive parameters can have the same number of values (e.g. ten). Consequently, a unique rating cannot be applied for all the parameters. Furthermore, strictly linear increase of importance for the descriptive values cannot be taken for granted (i.e. the ratings such as 1,3,4 and 6 can be expected, next to linear ones, e.g. 1,3,5,7). In that sense, the aggregation procedure introduced initially for determination of LI and SC factors has proved to be quite reliable (see next section).

The vast majority of relations between the parameters gathered in the LI and SC factors are not physically based, so the introduced procedure simply sums up the selected parameters values to provide the factors. The parameters' contributions to the factors were defined by their ratings and importance. However, the relation between the parameters that form the PA factor is physically based (Equation 5.4) so it should have been used in the assessment procedure. The relation was used for internal assessment, but not for the comparison between the sites. Establishing the absolute ranges for any more complex relation is difficult, unreliable and, probably, for the majority of cases worthless. Therefore, the same aggregation procedure was used as for the assessment of LI and SC factors, where the user can change the contribution of the parameters in assessing the PA and TE factors. If the contributions are modified, then the modified parameters' attributes should be used *consistently* for all the assessments conducted.

5.3.7 Testing aggregation procedure

This section contains some results of testing of the newly-introduced aggregation procedure. Since the section does not contain any facts that are necessary for understanding of the procedure, it can be skipped during the first reading.

The aggregation procedure is used to assess all the VAM factors, except for assessment of PA and TE internal ratings. The procedure was tested on the LI factor with respect to allocated ratings, confidence, threshold confidence and factor score.

The *ratings* introduced ranged from one to nine. In an ideal case, all the parameters would have the same number of values (e.g. five) and the same ratings within the specified range could be applied (e.g. 1, 3, 5, 7 and 9). If that was the case, the contribution of all the parameters to the factor score would be the same, presuming that the same value (e.g. lowest) is selected for each parameter. However, uniqueness of parameters with respect to number of values (or classes) cannot be expected. Rather then forcing uniqueness, the ratings should be introduced within approximately the same range for all the parameters. The ratings within the range need not to be exactly the same, because the 'confidence' is the main measure of their contribution to the factor score. Impact of the attribute 'confidence' on the score is explained below.

Confidence level is the probability of non-exceedance, assuming a normal distribution of the parameter. A confidence level or probability of 0.90, for example, yields a value of 1.645 standard deviations around the mean under the (non-standardised) normal distribution curve. The scores shown on a 0- to -10 scale of the User Interface are displayed with a 'standard deviation' s_n that presents non-confidence interval. The standard deviation s_n is obtained by dividing a half of the non-confidence interval (half of the scale) with the standard deviation that corresponds to defined probability. Graphically, this value defines the length of the horizontal bar left and right from the position of the rating at the scale (see Figure 5.2).

The impact of confidence level on the score was tested in several ways. Firstly, the same (uniform) confidence levels were assigned to all the parameters. Two sets of 'test input data' - minimum and maximum parameters ratings - were used in testing. The results of the testing (Figure 5.7) showed that the uniform confidence level produces no changes of the score, regardless of the parameter ratings introduced. As expected, the score confidence level also remains the same.

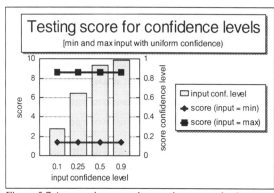

Figure 5.7 Aggregation procedure testing, example a)

If individual parameters have different confidence levels, the latter will influence the score. A score for a set of eight parameters was calculated where the parameters have moderate ratings (3 or 5) and various confidence levels (0.09, 0.18 and 0.36). The factor score obtained was approximately five. Figure 5.8 shows the way in which the various ratings and confidence levels (of one parameter) influence the score. If the parameter rating is five, it is in accordance with the overall score and it will more confirm than change the score. At the same time, the influence of the confidence level will be minimal. However, If the parameter rating differs from the score (e.g. the rating is one) it will cause a change in the score. *The change is then strongly dependent on the confidence level.* If the confidence level is higher (e.g. 0.36) the influence of the parameter will be higher. For example, (if the parameter rating is one) the score is much lower ('further away' from five) for the confidence level of 0.36 then for the confidence level of 0.09 (Figure 5.8).

Factor score The LI and SC factors could score between 1 and 9, i.e. between the minimum and maximum rating of individual parameters. All the parameters have the same or similar ratings (1,3,5,7,9 or 2,4,6,8). That does not mean, however, that all the parameters have the same importance in defining the score; for example, change in the rating of parameter 'Evidence..' has a stronger influence on the score than if same change occurs in the rating of parameter 'Precipitation'. Influence of the individual parameters could be increased by assigning considerably different ranges to the different parameters (e.g. to the parameter 'Precipitation' from 1 to 5 and to the parameter 'Evidence..' from 5 to 9). Nevertheless, it has been found more appropriate to keep approximately the same ratings and to introduce differences in the confidence level.

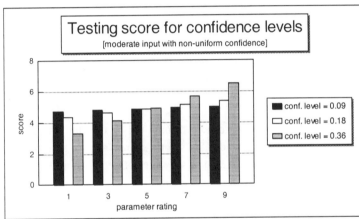

Figure 5.8 Aggregation procedure testing, example b)

As shown in Figure 5.8, if the rating of the parameter differs from the ratings of the majority of the other parameters, its contribution will be defined by the confidence level. In a situation like this the contribution of a parameter should be defined by its importance, so *the confidence level can be considered and used as a measure of importance rather then the measure of confidence in accuracy of value (rating) selected.* It is logical that parameters with higher confidence have a larger influence on the score. Therefore, this parameter attribute can still be presented to the user as 'confidence level' (as a matter of fact: statistically it is), with an explanation of its double functionality.

Confidence levels were assigned to the parameters and the score was tested for a number of combinations. Chosen confidence levels were shown in Tables 5.2 and 5.4 (Section 5.3). As a result of the testing, the score was divided into four ranges and descriptive factor ratings are assigned (Table 5.3, Section 5.3). Some results of the testing are shown in Figure 5.9, where a change in the score due to the change in rating of the parameter 'Evidence..' is examined. Three input sets were used for testing, containing maximum, minimum and average (moderate) ratings.

For the minimum input (minimum ratings of all the other parameters) the parameter 'Evidence..' has to have a rating of at least 3 in order to consider Likelihood of Release and Pollution - moderate (LI factor score: moderate). Further on, the LI factor will score 'high' (for the same input) only if the rating of the parameter 'Evidence..' is 9. It should be noted that this situation is extreme (mobility low, the best liner used, soil relatively impermeable, etc).

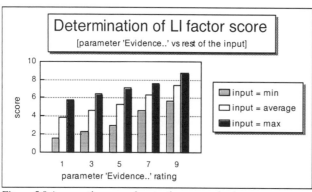

Figure 5.9 Aggregation procedure testing, example c)

At the other extreme situation (maximum ratings of all the other parameters) even without evidence of release ('Evidence..' rating is 1) Likelihood of release and pollution is moderate. With no evidence, but conviction ('Evidence..' rating is 3) the LI factor rating is 'high'. 'Evidence..' rating 5 (indirect evidence) gives already a very high Likelihood of release and aquifer pollution. For the moderate input set, the LI factor scores between extremes, and slight differences in individual ratings (especially mobility) can 'shift' the score to the neighbouring rating. Nevertheless, if indirect evidence of release exists, Likelihood of release and aquifer pollution will be 'high'.

Threshold confidence Reliability of the assessment can be secured only if a certain number of parameters are used for the assessment. The confidence of the factor score was used to define the threshold for acceptance of the assessment procedure. In this way, the defined threshold depends not only on number of parameters used, but also on their selection.

For example, an LI factor score should be calculated with the score confidence (p) of at least 0.50 (standard deviation s_n =0.741). That means that, on average, the confidence level of eight involved parameters should be at least 0.18 (s_n=2.19). The threshold value will also be reached with only five parameters if the parameter 'Evidence' is included, having the rating $\geq$ 5 (p=0.5, s_n=0.741), or having a lower rating, but if the parameter 'mobility' is included as well.

5.4 Software development and integration

The VAM has been developed for an MS Windows (NT 3.51) environment by using object-orientated Borland Delphi Developer, a tool based on Object Pascal as a programming language.

Most of the requirements posed during VAM development were of a standard type: the software architecture and user interface should be kept as simple as possible; the tasks (module operations) clearly defined; content-sensitive on-line help as well as operation control provided, etc. Special attention was paid to software transparency, allowing the user an insight into all major operations that the software performs. This is closely related to the 'adaptability', a rather uncommon software feature that permits the user to modify VAM parameters and their attributes.

The simplified software algorithm (Figure 5.10) gives an overview of the main operations that are carried out during the vulnerability assessment. The algorithm is accompanied by Table 5.6

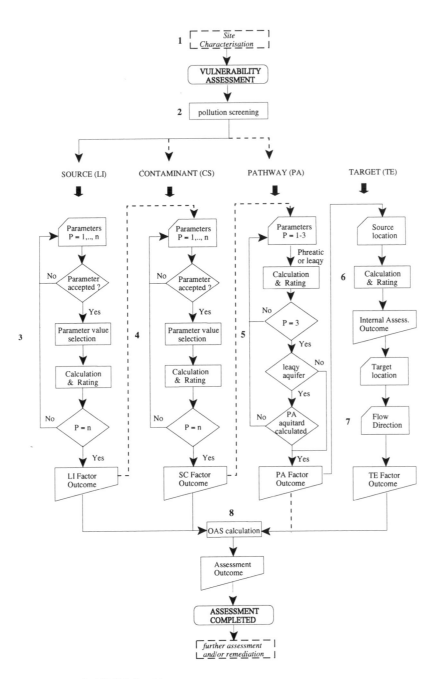

Figure 5.10   Simplified VAM algorithm

that relates the main operations, involved factors and parameters, and the DSS components utilised in performing the operations.

Table 5.6 Overview of main VAM operations

OPERATION	FACTOR / PARAMETER	OPERATION TAKEN IN	REMARK
1.REVIEW, REVISION	mobility, area, indicator parameter (Koc), conductivity map, groundwater level map (gradients)	VAM, GIS (review); SCM, GIS, HDB (revision)	review and possible revision of Site Characterisation results
2.SELECTION DEFINITION	contaminant source type, aquifer type	VAM	groundwater pollution screening
3.INPUT SELECTION CALCULATION . RATING	LI factor & parameters	VAM	assessment of the Likelihood of release and aquifer pollution (outcome: descriptive rating)
4.ditto 3	SC factor & parameters	VAM	assessment of the contaminant severity (outcome: descriptive rating)
5. ditto 3	PA factor & parameters: hydraulic conductivity, k hydraulic gradient, I retardation R: - porosity, n - Koc - org. carbon content, Foc - dry bulk density, Pb	VAM, HDB, GIS, SCM VAM, HDB, GIS, SCM VAM, GIS, SCM VAM, HDB, GIS, SCM VAM, CDB, GIS, SCM VAM, HDB, GIS, SCM VAM, HDB, GIS, SCM	assessment of the pathway activity, resulting in internal and external rating maps for the parameters k, I , R and for PA factor as a whole.
6.INPUT SELECTION CALCULATION	PA factor (internal rating), groundwater level map (flow direction)	VAM, GIS	internal assessment of the targets exposure (outcome: Potential Hazard Map, Pathway Map and Distance Map)
7. ditto 3	PA factor parameters (external rating) distance, flow direction	VAM	assessment of the targets exposure for site comparison (outcome: descriptive rating)
8.CALCULATION	LI, SC, PA and TE factors	VAM	calculation of OAS (outcome: descriptive rating)

6 - operation numbers refer to the numbers given in Fig 5.10; SCM - Site Characterisation Module; HDB - Hydrogeological DataBase; CDB - Chemical DataBase; GIS - Geographical Information System;

Although the VAM can without any modification be used as an independent piece of software, it is primarily considered as an integral part of the DSS. Therefore the user is asked first to retrieve a Site Characterisation (SC) results file, if available . If the file is not available, the assessment can still be conducted partially (i.e. assessment of LI and SC factors) or completely (if the hydraulic conductivity and groundwater level data are available). The SC result file (Figure 5.11) provides the user with some basic information on the investigated site (source area, contaminant mobility, conductivity map, etc.) that are needed for estimation of VAM factors and parameters. The user can not only review, but also revise the SC results by activating the SCM (operation 1 in Table 5.6 and Figure 5.10).

The same User Interface window (Figure 5.11) contains a procedure for groundwater pollution screening (LeGrand, 1980) based on a contaminant source type and aquifer type. The contaminant source type is obtained from the SC result file, while the user has to select an aquifer type.[10]

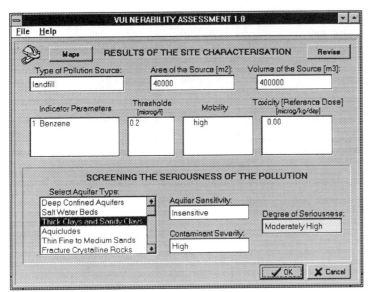

Figure 5.11 SC result file

After the SC result file is reviewed/revised and screening conducted, the user can perform VA (operations 3-8) by utilizing various software applications integrated in the DSS (Table 5.6). The parameter attributes, together with defaults, are placed in the local VAM (Paradox) database, while the field data are stored in the REGIS Hydrogeological Data Base (HDB). Dynamic Data Exchange is establish with the Chemical Data Base (CDB) for selection of organic carbon/water partition coefficient K_{oc}. A REGIS GIS is used to create and display maps. Eventually, the qualitative information (knowledge) on parameters (i.e. definitions, importance, methods, references, etc) can be obtained from the Site Characterisation Module (SCM).

Adaptability must be considered as the important feature for any software that contains a rating-based methodology. Once encapsulated, the methodology is sometimes not used because of disagreement on parameters included and their relative importance. Therefore the VAM provides to the user with an option to modify parameters and their attributes. Moreover, a new parameter can be added to the LI and SC factors. Alternatively the majority of parameters can be

---

[10] Each source type is (in the SCM) associated with contaminants most commonly related to the type. That practically means that the severity of contaminants is used for pollution screening.

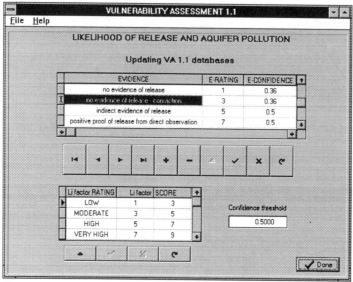

Figure 5.12 Access to VAM databases (an example)

disregarded. By clicking at the button 'unlock' (see Figure 5.2, for an example) the user is prompted to the type of window shown in Figure 5.12. The ratings and confidence levels for all included factors can also be adapted (improved) by the user. Thanks to this adaptability, the VAM can be called 'interactive' in the real sense of the term. It interacts with the user by transferring the encapsulated knowledge on vulnerability assessment. Moreover, the user has the opportunity to transfer her/his knowledge into the computer, and improve the encapsulated assessment methodology.

5.5 Case study

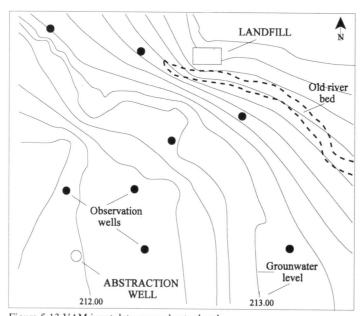

Figure 5.13 VAM input data: groundwater levels

A landfill located in the north-eastern part of the investigated area (Figure 5.13) has been in operation since 1988, covering an area of approximately 0.4 ha. The landfill is equipped with a geomembrane liner and drainage system, but no treatment of collected water is provided. Groundwater has been abstracted about 2 km south-west from the landfill and used by local farmers for domestic purposes. Several observation wells are situated between the landfill and the abstraction well as a part of the regional groundwater monitoring network.

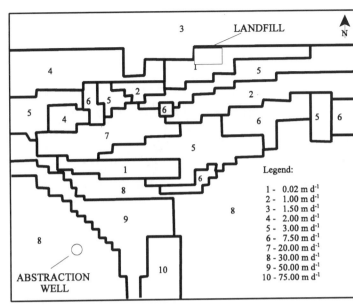

Figure 5.14 VAM input data: hydraulic conductivity

During the last wet season, groundwater level rose above the bottom of an old river bed, showing the presence of some oil derivative in the water. Samples taken on regular basis from the observation wells showed, however, no traces of contaminants.

The Water Authority decided to carry out hydrogeological investigations in order to assess the groundwater pollution situation. The results of the assessment are to be compared with results obtained from other investigated sites in order to establish a priority list for possible remediation measures. In addition, if the situation at the site appears to be unsatisfactory in terms of pollution hazard for the abstraction well, the best alternative well location has to be found.

The results of field, laboratory and previous desk-top investigations were gathered and checked during the Site Characterisation; therefrom, a representative groundwater level map and hydraulic conductivity map were prepared (Figures 5.13 and 5.14, respectively), forming the basis for the VA.

Table 5.6 Selected values for LI and SC parameters

LI FACTOR		CS FACTOR	
PARAMETER	**SELECTED VALUE**	**PARAMETER**	**SELECTED VALUE**
evidence..	Indirect evidence	toxicity	SAX level 2 or NFPA level 2
soil permeability	relatively permeable	radioactivity	At or below background level
mobility	high (indicator parameter - benzene)	persistence	Substituted and other ring compounds
depth to aquifer	5-20 m	mobility	high
precipitation	400-800 mm/y	quantity	1000 - 2000 t

Results of value selection for LI and CS parameters are presented in Table 5.6. It was assumed that a contaminant has already been released from the landfill. Therefore parameters Liner and Leachate Collection are not taken in account. Parameter Corrosiveness is not found relevant to the assessment. The LI factor was found to be 'high', whereas the severity of the indicator parameter (benzene) is assessed as 'moderate'.

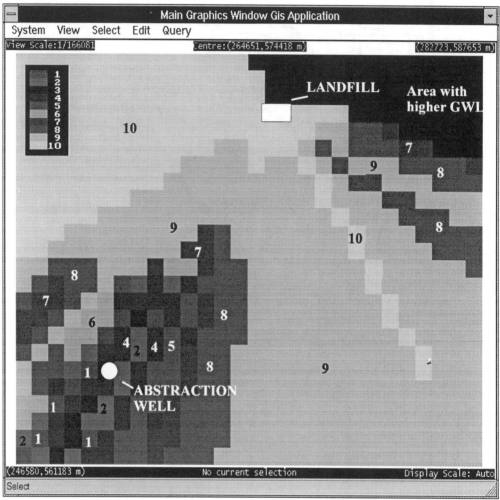

Figure 5.15 Potential Hazard Map (PHM)

A number of maps (Figures 5.15, 5.16 and 5.17) have been prepared following the procedure outlined in Section 5.3. According to the Potential Hazard Map (Figure 5.15) the location of the abstraction well was relatively satisfactory (comparing with other possible well-target locations). A more favourable locations can hardly be chosen under present conditions (source location and PA distribution) at the site. As expected, the south-east part of investigated area scored quite high, owing to high PA values that neutralise the favourable effect of increasing distance. The north-west region also scored high, although the VA values were somewhat lower than in the South-East. That was primarily due to the gradients being the highest in this region. Besides, transport of pollutant towards this region would not be affected by any of the low conductivity areas (one of them is located immediately south of the source). This is however true only if the assumption of linearity holds. The Grounwater Pathway Map (Figure 5.17) indicates that the main contaminant load would be transported north of the abstraction well. Considering the approximative nature of the assessment, no guarantees can, however, be given that abstracted water would remain unpolluted.

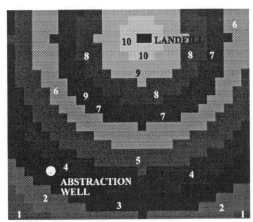

Figure 5.16 Distance Map   Figure 5.17 Groundwater Pathway Map

Finally, the OAS was estimated for the purpose of establishing a priority list. The PA between the landfill and abstraction well was estimated as 'moderate' (external rating), the same as the TE (a flow direction is 'possibly towards the target'). The OAS was determined as 'moderate'. A recommendation was issued regarding taking additional water samples from the monitoring wells.

6. KNOWLEDGE-BASED MODULE FOR GROUNDWATER POLLUTION MODELLING

6.1 Introduction

Numerical models have become irreplaceable tools used in the struggle to solve groundwater pollution problems. Despite modelcodes becoming more and more sophisticated, the modelling process is still heavily subject to errors. The latter originate from various sources (e.g. limitations of encoded technique and modelcode errors), but most of them are made by a modeller. Curiously enough, some of the most serious errors are made in those steps of the modelling process that precede or follow actual use of modelcode. Advanced pre- and post-processors have been developed (and sometimes coupled with databases) reducing (numerical) input errors and alleviating interpretation of modelling results. Errors are, however, still made, because modelling, i.e. processing of numerical (quantitative) information (represented in the model as geometry, parameters and variables), requires also knowledge (qualitative information) on processed numerical information, modelling procedure, modelcode and encoded technique.

There are very few (but valuable) examples of encapsulation and integration of knowledge that would provide assistance in groundwater modelling (see overview, Section 2.2). Lack of formalised expertise has made application of rule-based knowledge representation forms quite limited. That was probably one of the reasons to use a hypertext environment for both knowledge encapsulation and as a DSS platform (Newell et al, 1990, Rifai et al, 1994). These 'hypertext-based' DSSs, however, lack the usual GIS graphical capabilities. On the other hand, knowledge encapsulated (to-date) in GIS-based DSSs (e.g. Fedra, 1993) is mainly related to environmental management, surface water modelling or a study area specifics.

Responding to the perceived need for encapsulation of knowledge on groundwater modelling, a prototype of a knowledge base (called GMM) has been developed to support modelling of point-source pollution problems. The GMM contains general knowledge on the major modelling steps, organised in an 'electronic' Modelling Protocol (MP). At this stage, special attention has been paid to the first two steps of the MP where the purpose of modelling and the conceptual model have to be determined. These steps are crucial for the model design, when actual application of modelcode begins. Attention has also been paid to the reporting on modelling results; since optimal (re)use of modelling results asks for proper documenting (reporting), GMM contains recommendations on the contents of modelling reports, while REPORTER (another DSS component) provides a predefined reporting format, in order to both speed up reporting and secure consistency of reports.

Beside the MP part, the GMM also contains three additional (independent) software applications, namely MCM, DE1 and RE1. These applications provide a particular assistance in determination of general model complexity, dispersivity parameters and retardation factor, respectively. DE1 and RE1 are used by other DSS knowledge based modules during the site characterisation and/or vulnerability assessment.

The GMM is one of the Knowledge-Based Modules of a DSS for Groundwater Pollution Assessment. It was integrated with REGIS and MODFLOW groundwater modelcode. REGIS was coupled with MODFLOW through the PMWIN modelling environment and through the REMOD coupling module (see also Section 3.4).

GMM content and organisation will be described in the following three section of this chapter. The first steps in defining the module content were analysis of groundwater modelling process (modelling tasks) and establishment of modelling protocol (Section 6.2). That allowed (subsequently performed) acquisition (and systematisation) of knowledge required for each step of the modelling procedure (Section 6.3). Systematisation and formalisation related to representation of knowledge in electronic form will be briefly discussed in Section 6.4. It can be noticed that GMM content and organisation mirror the first four steps of the encapsulation process (i.e. definition of the problem, knowledge acquisition, systematisation and formalisation), as given in Section 3.5. The last two encapsulation steps, software development and integration, will be described in Section 6.5.

6.2 Definition and ordering of modelling tasks

The groundwater modelling process (or rather a 'procedure' - indicating process complexity) consists of several steps that are associated with particular modelling tasks. For modelling (like any other random) procedure, it holds that the content of the tasks and their place (position) within the procedure need to be clearly defined. Therefore, definition and ordering of modelling tasks will be addressed here in order to establish a general modelling protocol or 'assessment framework for groundwater modelling'.

The purpose of the modelling, nature of the groundwater (pollution) problems and data availability are the main factors that define the general content of the modelling procedure. These factors are very variable, so that no universal modelling protocol can be established. Even if only major modelling steps are included in the general protocol, some of them might be optional, so that professional knowledge needs to be utilised while applying the protocol; for example:

- if a model is used for interpretative purposes, prediction (and sometimes calibration) step(s) will be left out;
- if the intention is to develop a generic model, no information from a particular site is required;
- if a second set of field data does not exist, model verification has to be skipped;
- model predictions can be compared to the actual outcome in field conditions (postaudit) only in exceptional circumstances, etc.

Nevertheless, the general modelling protocol should contain all major modelling steps. It can be subsequently worked out in detail for each (group of) particular problem(s) and for each type of modelling application. According to the purposes of the GMM (and DSS for Groundwater Pollution Assessment), the modelling protocol (to be encoded in the GMM) will be established for (predictive) modelling of local-source pollution problems in an intergranular environment.[1] Despite this limitation, the GMM protocol can be very useful for setting up protocol(s) for other problems, as well as other types of modelling applications, because the core of the modelling procedure is the same. Moreover, the GMM could be used in other situations (the large portion of information required for modelling is the same), if the user is aware of differences between problems and differences between types of modelling applications.

Several documents have been found that contain a description of general modelling protocol. Various (versions of the) protocol(s) were examined and compared (Table 6.1), resulting in the protocol (for predictive modelling of point-source pollution problems) that will be used in the GMM ('GMM protocol').

Modelling steps (tasks) are carried out basically in sequential order, but involving iteration of some sequences as well as the modelling procedure as a whole (modelling is analogous to the scientific-iterative method of formulating a hypothesis and testing it). A GMM protocol flow chart is given in Figure 6.1. The protocol given by Anderson and Woessner (Table 6.1) contains a verification loop inside the Code Selection step and a revision loop that connects the Postaudit and Conceptual Model. According to Bear et al (Table 6.1), the results of the calibration step could lead to (additional) field data collection and revision of Model Conceptualisation. ASTM has introduced two decision nodes in the protocol, similar to the nodes 1 and 2 given in Figure 6.1 (however, the possible revision reaches only Conceptual Model, and not Study Objective). Finally, there is almost complete agreement between decision nodes (and thus possible iterations) given in the OSWER and GMM protocols.

The comparison presented in Table 6.1 showed a high degree of similarity in content of the

---

[1] Anderson and Woessner (1992) divided modelling applications into three basic types: generic, interpretative and predictive. A clear distinction between the predictive and interpretive models can be made in the sense that the interpretative models do not include prediction; predictive models, however, include not only simulation of future GWS behaviour, but also simulation of present state of the system (otherwise a prediction could not be made). In order to simulate properly the current state of the system, the flow system analysis and sensitivity analysis are (should be) carried out. Modelling applications dedicated to these kinds of analysis (and not to prediction) are considered as interpretative applications. Therefore, predictive model applications are (partially) interpretative as well.

The generic type is deliberately 'excluded' from the modelling protocol in order to point out that the GMM protocol includes more modelling steps than a generic application usually entails (the same holds for knowledge encapsulated into the GMM).

Other classifications of modelling applications will be discussed in the section 'Knowledge acquisition'.

Table 6.1 Various views on general modelling protocol

GMM protocol	Anderson and Woessner (1992)	Bear et al (1992)	OSWER (1994)	ASTM (1993a)
1 Define Purpose	Define purpose	Formulation of objectives	Establish modelling objectives	Study objectives
			Establish project management plan (also 7.)	
2 Model Conceptualisation	Conceptual model	Available data review and interpretation	Collect, organise and interpret available data	Conceptual model
		Model Conceptualisation	Prepare a conceptual model	
3 Modelcode Selection	Mathematical model	Code selection	Select a suitable mathematical code	Computer code selection
	Code selection			
4 Model Design	Model design	Field data collection	Set up the model and perform input estimation	Model construction
		Input data preparation		
5 Model Calibration (including sensitivity analysis	Calibration	Calibration and sensitivity analysis	History matching (calibration using field data) including calibration sensitivity analysis	Calibration and sensitivity analysis
	Verification			
6 Model Prediction (including sensitivity analysis	Prediction (including sensitivity analysis)	Predictive runs	simulate the scenarios (including predictive sensitivity analysis)	Predictive simulations
		Uncertainty analysis		
7 Report Completion (reporting is a part of each modelling step)	Presentation of results		Evaluate the overall effectiveness	Documenting modelling study
8 Postaudit	Postaudit		Postsimulation analysis	Postaudit

protocol(s) suggested by various authors. The main differences were encountered in treatment of (available) data, reporting and possible iterations. These issues called for more attention and will be discussed below.

Field Information and Modelling Procedure

OSWER (1994) considers 'collection, organisation and interpretation of available data' as an independent modelling step that precedes Model Conceptualisation. Similarly, 'review and

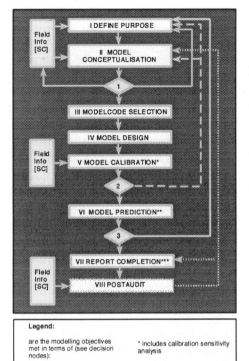

Figure 6.1 Modelling Protocol

interpretation of available data' is included in the modelling protocol of Bear et al (1992). The same authors suggest (additional?) field data collection prior input data preparation (see Table 6.1). Anderson and Woessner (1992) state that data should be assembled during the Model Conceptualisation. In all these approaches 'available information' is considered after the purpose of modelling is defined (only OSWER suggests possible revision of the modelling purpose (i.e. modelling objectives), after available information is collected, organised and interpreted). However, no (even preliminary) modelling purpose can be defined without some knowledge on the actual problem! At least some (but usually most of) the 'field' data required for the modelling are already available prior to the modelling procedure taking place; it is usually said that these data are available from 'regional hydrogeological mapping' and 'previous investigations'. Nevertheless, characterisation of the pollution site is the first task to be carried out when dealing with point-source groundwater pollution problems. In other words, the Site Characterisation (SC) precedes (being a prerequisite for) any other assessment (including modelling) or remediation of groundwater pollution problems. Accordingly, the (processed) field information (rather then field data) required for modelling with the GMM is actually an outcome of SC. The SC is performed with assistance of the Site Characterisation Module (Chapter 4).

Field information is needed several times during the modelling procedure (see Figure 6.1). Information use in Model Calibration (history matching) and Postaudit ('implemented scenario' matching) is straightforward. Define Purpose (DP) and Model Conceptualisation (MC) are the steps where the SC outcome is of particular importance; these modelling steps require a summary of and the detailed outcome of SC, respectively. Although the modelling procedure formally begins with definition of the purpose of modelling, the summary of the SC outcome should be available a priori.[2] The SC summary should contain concise information on general setting (geographical position of the investigated area, topography, climate, hydrology, geology, hydrogeology, etc) and the pollution situation (contaminant characteristics, spreading, potential

---

[2] This is regardless of the level at which SC has been conducted (from general to very detailed), and regardless of the quality of the results; if available knowledge is, for instance, very poor, a decision can be made (while defining the purpose of modelling) not to conduct modelling at all - i.e. no purpose.

targets, etc). Additionally, the regional (overview) map should be made available, next to the overview of information sources used. Unlike Define Purpose (where only the SC summary is required), Model Conceptualisation is seen as a step where the (detailed) SC outcome is extensively used.

Reporting on Modelling Procedure

It is very important to report on each modelling steps during step execution, or immediately after. By doing that, the user obtains a clear overview of decisions made and results obtained, both needed for the following modelling step(s). The track of the modelling procedure should be firmly kept, and immediate reporting is the best way of doing that. For example, the impact of changes (introduced due to iterativeness in procedure) will be more clear and easier to explain if the previous state (before changes took place) is already documented. A modelling knowledge base should provide the user with advice on content of the report, that would, together with the predefined reporting format (available in a text processor), alleviate the reporting process and contribute to report consistency.

Documenting a modelling study is usually mentioned as (one of) the last modelling steps (Table 6.1). OSWER suggested 'establishing of the project management plan' already at the beginning of the modelling procedure. The plan contains not only budget estimation and organisation of the modelling team, but also establishment of documenting procedure 'up front to assure that an independent reviewer can duplicate the modelling results or perform a post-application assessment using the documentation'. The documenting procedure is actually a list of items that need to be addressed and documented during the modelling procedure.

The report, being subjected to changes, should be prepared in draft form, but no item should be left out (forgotten). Reporting on some items can be done only at the end of the modelling procedure ('evaluation of overall effectiveness'), so the report in its final form is prepared during the Report Completion, the last modelling step (if Postaudit is not taken into account).

Iterativeness in Modelling Procedure

Whether some modelling steps (or groups of modelling steps) need to be repeated depends on decisions taken during the modelling procedure. The decisions that can cause major iterations are shown as 'decision nodes' in Figure 6.1. The outcome of the Model Conceptualisation (decision node 1) could lead to:

- additional SC in order to obtain additional information required for model conceptualisation;
- revision of the modelling objectives in Define Purpose (for instance, if it appears (during the conceptualisation) that the posed objectives are too ambitious),
- the third step of the modelling protocol (Modelcode Selection).

If calibration targets are not met (decision node 2), the conceptual model needs to be improved, or modelling objectives reconsidered. Often, during calibration, it becomes apparent that there are no realistic values of the hydraulic properties which will allow the model to reproduce the calibration targets. In these cases the conceptual model of the site may need to be revised, bringing, consequently, changes in model construction. In addition, the source and quality of data used to establish calibration targets may need to be re-examined. In some cases, the only option left is revision of modelling objectives.

The modelling objectives have to be revised (decision node 3) if simulated scenarios failed to provide required and expected information. This decision is primarily based on results of the predictive modelling and related sensitivity analysis.[3]

The results of the postaudit should be used to improve the model (if a budget is available), or at least added to the modelling report.

Beside the iterations shown in Figure 6.1, the modelling procedure contains a number of 'local' (but not less important) iterations, such those related to code verification and model calibration.

6.3 Knowledge acquisition

General knowledge required for modelling was acquired from books, reports, articles, Internet, CD-ROMs and though personal communication. Some (parts of the) acquired documents were scanned and included in the GMM without any alternations (e.g. ASTM guides). Since undue complexity, mis-conceptualisation and misinterpretation are (often-made) serious modelling errors, model complexity, model conceptualisation and reporting received special attention during the knowledge acquisition. Accordingly, these issues will be discussed below in more detail. Only the basic content of the other modelling steps will be given in this section. Extensive description, including scanned documents, is available from the GMM.

6.3.1 Define Purpose (DP)

A clear definition of modelling purpose at the outset of the study is crucial because at that stage decisions on the necessity for modelling and on model global complexity have to be taken. The modeller should remain aware of the modelling purpose at each stage of the modelling procedure because that purpose defines or influences all the other steps, especially Model Conceptualisation and Model Design.

---

[3] Some issues addressed in modelling steps 'Postsimulation Analysis' and 'Evaluation of Overall Effectiveness' (OSWER, 1994) should also be considered here, such as: 'the model application should provide information being sought by management for decision making' or 'the model application results should be acceptable to all relevant parties'. These, and similar issues, however, require further elaboration.

The principal input information is the Site Characterisation (SC) summary. In the context of the DSS, additional input can be the results of Vulnerability Assessment (VA), if performed.[4] The DP task includes definition of the following main items: MDO, Modelling Objectives (MO) and global Model Complexity (MCO).

Management Decision Objectives (MDO) are defined as 'the information needs required to identify courses of action necessary for reaching environmental and regulatory goals'. MDO should be clearly specified initially, because defining objectives of a modelling study (modelling objectives) is primarily a management problem.

The management of any system means making decisions aimed at achieving the system's goals without violating specified technical and nontechnical constraints imposed upon it. In a groundwater system, management decisions may be related to rates and locations of pumping or recharge wells, changes in water quality, pump-and-treat operations, etc. The management decision objectives (i.e. the values of the management's objective function) should be set up in such a way as to minimise costs and adverse effects, and to maximise benefits. The MDO are defined in terms of decision variables (e.g. areal and temporal distribution of pumpage), and decision constraints. The constraints can be divided into: (1) technical constraints (specifics of GWS, contaminants and management practice at the site), and (2) nontechnical constraints (available resources and legal and regulatory framework). The constraints are eventually expressed in terms of decision variables (future values of state variables or hydrological stresses: groundwater level, concentration of specific contaminants in the water, pumping rates, etc.).

Modelling Objectives (MO) may be defined as 'the information that the modelling application is expected to provide so that management can evaluate potential courses of action'. The role of a modelling study in the pursuit of MDO should be established, i.e. MDO should be translated into MO. The following questions should be answered (jointly by the manager and the modeller):

- 'what kind of information is expected from the results of modelling study'? (The answer on this question defines MO.) and

- 'is the modelling the best way (considering constraints) to provide that information'? (The answer on this question also defines the necessity for modelling: should a modelling study be carried or not?) [5]

---

[4] Management Decision Objectives (MDO) could also be considered as DP input. However, they are preferably considered as a part of the DP task; if MDO are not (clearly) defined prior to implementing modelling procedure, then that has to be done during the DP step. Unlike MDO, SC and VA results are outcomes of activities that might be carried out independently of groundwater modelling.

[5] It may be that a model is not necessary after all, and that the problem at the heart of the study could be solved using different approach; or it may be that a single analytical model can provide answer without numerical modelling. McLaughlin and Johnson (1987) showed, for example, that all of the substantial results obtained in three

(continued...)

MO should be expressed (as much as possible) in the terms of model parameters and variables. OSWER wrote that: 'MO should be expressed, for example, in the terms of calibration targets and prediction period'. Precise definition of a calibration target already at this modelling stage could be possible if a refinement of already existing modelling study is planned; otherwise only a rough specification can be given. MO should be based primarily upon existing information about the specifics of the GWS, contaminants and management practice at the site (thus 'technical constraints' but now applied to modelling decisions). OSWER suggests that 'the purpose of the model application (e.g. data organisation, understanding of the system, planning additional field characterisation, prediction, or evaluation of remediation alternatives) should be defined during the development of the modelling objectives'[6].

In the protocol, 'define purpose' should mean not only definition of a type of modelling application, but also (as precisely as possible) specification of MO. OSWER suggests that 'the purpose of the model application should be reviewed during the course of the project and, if necessary, modified'. As far as modelling application types are concerned, this is not so likely to happen: for instance, if the model is developed for evaluation of remediation alternatives, what can be an alternative purpose? (An imaginable situation might be that Model Conceptualisation and/or further steps indicate model prediction as not reliable/feasible, so that the model is unwillingly altered from 'prediction' to 'understanding of the system'.) However, if the purpose is specified at a much more detailed level (in terms of MO), then modification of the purpose (i.e. MO) can be expected (for example, in the terms of calibration targets and prediction period).

The importance of decisions which will be influenced by the model results should be considered while setting up the MO; that also includes estimation of sensitivity of those decisions to the range of possible or likely modelling outcomes. Findings on these issues should be reported (see Reporting), together with all assumptions incorporated within the MO (the assumptions should be reviewed with respect to reality and their potential impacts on MDO).

Model Complexity (MCO) The term 'Model Complexity' is commonly used, but not yet

---

[5](...continued)
modelling studies could have been anticipated with a few simple calculations using the Theis drawdown solution. Use of desk-top analysis (budget calculation, groundwater level analysis, drawdown computations) is therefore strongly recommended to determine first whether a modelling study is necessary, and then how it should be conducted.

[6] This is, however, still quite a broad definition of purpose, that covers both interpretative and predictive types of modelling application. The general protocol could be further worked out for each of specific applications, which would not be so difficult at this level of generality. A few steps would be used for 'understanding of system' or 'planning additional field characterisation' (as already mentioned, a sort of interpretative model is a prerequisite for predictive one). Besides, evaluation of remediation alternatives also indicates prediction; this can be expanded even further (e.g. pump-and-treat systems, containment systems, 'no action' alternatives), having as a consequence a set of specific protocols.

precisely defined.[7] Decisions related to MCO are taken during several modelling steps, as illustrated below:

1. The first modelling step (DP) provides a framework for MCO; this preliminary specification of MCO is based on the SC summary and MO.
2. Decisions on MCO are also taken during the MC. For example, hydrogeological units are combined, disregarded, or included in a model as independent units. The next example concerns anisotropy, which, if recognised, requires estimation of directional hydraulic conductivity. The results of MC could lead to revision of MO and MCO.
3. Modelcode Selection is based on (required) Model Complexity. In practice, however, model availability can affect MCO.
4. Decisions on (for example) cell-size and time-step taken during the Model Design also define Model Complexity.
5. Model Calibration could show the necessity for revision of MO and MCO.

Once defined, the purpose of the modelling application is subjected to changes, due to the iterative nature of the modelling procedure. The main 'decision nodes' where decisions on MO and MCO revision can be taken, are shown in Figure 6.1.

6.3.2 Model Conceptualisation (MC)

Providing assistance during the Model Conceptualisation is the one of the main purposes of the GMM. It is assumed that Site Characterisation is conducted with assistance of the SCM and that results are stored in REGIS. If the results of SC are insufficient, additional field investigations have to take place (see Figure 6.1). MC is carried out according to the DP outcome, expressed through MO and MCO. Results of the MC should be immediately reported according to the instructions given in the Reporter. MC yields information required for Model Selection (MSEL) and Model Design (MD).

Model Conceptualisation consists of two (sets of) tasks:

---

[7] Decisions on model complexity taken during the first modelling step (DP) are of a general type ('framework for model complexity'). As these decisions set up, at the same time, the framework for the MC, the relation between MCO and MC could be questioned. The possible answers are:

- MC already begins during the DP (however, DP and MC are, and should be, treated as independent steps), or
- MCO defines MC (but also the other way around, because a decision on Model Complexity taken during Model Design is based on MC), or
- MCO and MC are synonyms (however MCO is not defined only during the MC);

In order to resolve this issue Model Conceptualisation should be defined essentially as '*interpretation and transformation of results of Site Characterisation for the purpose of mathematical modelling*'. Then the (determination of) Model Complexity can be seen as a process much broader than Model Conceptualisation.

1. interpretation of results of Site Characterisation for the purpose of mathematical modelling (basically, selection of information according to the MO and MCO directives and its presentation in a form required for Model Design). This is the essential MC task.

2. further (specific) conceptualisation for the modelling purpose (heterogeneity, anisotropy, etc). This is also a step in determination of Model Complexity.

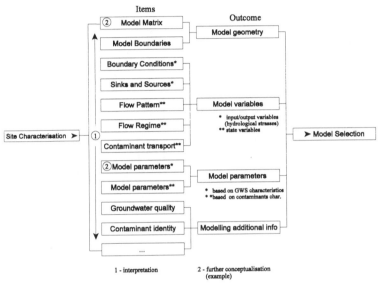

Figure 6.2 Model Conceptualisation items

The main difficulty while ordering information required for MC is that decisions (that have to be taken during MC) are usually based on several pieces of information that should be consider simultaneously. For instance, model boundaries are determined by using information on hydrogeological units, groundwater with-drawal/recharge, surface water courses, flow pattern (groundwater head), location of pollution source and potential targets, etc. Moreover, the factors involved are not (and cannot be) defined with the same accuracy; they also do not have the same importance, which can be, in addition, influenced by the MDO. Nevertheless, availability of SC results (SCM output) in a proper form (stored in a database with predefined options for spatial and temporal data presentation) could substantially alleviate MC. The relation between SC outcome and the Model Conceptualisation (first MC task) is given in Table 6.2. A clear distinction should be made between the conceptual model of GWS and Model Conceptualisation. The conceptual model of the GWS is developed independently of the mathematical model, showing a geohydrologist's concept of physical GWS. During the Model Conceptualisation the GWS model is interpreted and further conceptualised for the purpose of mathematical simulation. The items that are the subject of Model Conceptualisation (further in the text: MC items) and MC outcomes are presented in Figure 6.2. In short, SC outcome is used by MC items (components) in order to define MC outcome.[8]

---

[8] Organisation of SC parameters is carried out according to the information required to develop a conceptual model of the GWS and to estimate pollution in the system (see also Footnote 5 on page 54).

Table 6.2 Relation between Site Characterisation and Model Conceptualisation

Results of SC given as fulfilment of SC objectives	SC Output	SC Output form - MC Input sources	Relevance for Model conceptualisation item (for development of mathematical model)
1. CONCEPTUAL (physical) GWS MODEL	Integrated information on groundwater system where pollution (might) take(s) place	See the individual items (1.1 - 1.3)	See the individual items (1.1 - 1.3)
1.1 System matrix 1.1.1 Rock type 1.1.2 Texture 1.1.3 Colour 1.1.4 Structures 1.1.5 Spatial distribution	Principal output: spatial distribution of (lithostratigraphic, and after translation) hydrogeological units; Additional output: optional	Principal output form: - maps: 'thickness' of **geological** (lithostratigraphic) units and map 'depth to' showing bottom unit - cross-sections: **lithostratigraphy** - maps: 'thickness' of **hydrogeological** units and map 'depth to' showing bottom unit - cross-sections: **hydrogeology** - table: chronostratigraphy, lithology, hydrogeological units Additional output form: - map: **structures** and their characteristics - diagram: **texture-triangle**	Principal use: - **Model matrix** development contain interpretation (and additional conceptualisation) of GWS for (mathematical) modelling purposes (hydrogeological units in the model and their spatial distribution). Additional use: - **Model Boundaries**, especially while setting up some of physical boundaries (i.e. impermeable rocks or geological structures that act as a hydrogeological barrier). This task should be conducted when some other SC outputs become available (e.g. groundwater head, surface courses, etc). - Indirectly, the System Matrix output is also used during determination of **Boundary Conditions** (precisely: 'no flow' boundaries) and **Flow Pattern** (precisely: recharge/discharge pattern).
1.2 System boundaries & boundary conditions.	Principal output: spatial and temporal distribution of GWS inflows and outflows required for setting up model boundaries and boundary conditions	Principal output form: - GW '**budget**' table with elements of GW budget and results of the budget calculation [if not steady state (ie transient) flow is considered, the budget should be calculated for several representative time intervals (seasonal, new pumping rates, etc)] see comments - various maps (see individual items)	Principal use (for details see the individual items): - **Model Boundaries,** - **Boundary Conditions** - **Flow Pattern and Flow Regime** - **Sinks and Sources** Additional use: - **Contaminant transport**

Table 6.2 continued

Results of SC given as fulfilment of SC objectives	SC Output	SC Output form - MC Input sources	Relevance for Model conceptualisation item (for development of mathematical model)
1.2.1 Precipitation	precipitation: temporal and spatial distribution	- numerical output in the budget table - (if not steady state) chart and table showing temporal distribution of **precipitation** - (if not uniform) **hydrology I** map with spatial distribution of precipitation	- **Boundary Conditions** - precipitation and evapo(transpi)ration are principal factor that define net-recharge, being boundary condition at the upper (top) model boundary (influence of unsaturated zone is not considered here). - **Flow Regime** - charts and tables that show temporal distribution of precipitation and evaporation are needed for understanding of the flow regime in the system.
1.2.2 Evaporation	evaporation: temporal and spatial distribution	- numerical output in the budget table - (if not steady state) chart and table showing temporal (mind time step!) distribution of **evaporation** - (if not uniform) **hydrology I** map with spatial distribution of evaporation	- ditto above
1.2.3 Groundwater level (head)	GW level\head: temporal and spatial distribution	- **GWL\head** map for each aquifer - (if not steady state) hydrograph showing temporal (mind time step!) distribution	Principal use: - **Flow Pattern** - the GW level (head) hydroisohypses exhibit a certain pattern that is used in determination of general direction of the groundwater flow through the system. The GW level (head) map could also indicate GWS recharge and discharge zones, as well as relation between aquifers (seepage-infiltration zones) - **Model Boundaries** that do not resemble aquifer boundaries are usually set up along the hydroisohypse (i.e. perpendicular on streamline - flow boundary) or perpendicular on hydroisohypse (along the streamline - no flow boundary) - **Boundary Conditions** - initial estimation of inflow/outflow at boundaries placed along the streamlines - **Flow Regime** - hydrograph that shows temporal distribution of GW level (head) is needed for understanding of flow regime in the system. Additional use: - **Sinks and Sources** - GW level map indicates impact of recharge/discharge wells on GWS and interaction between GWS and surface water courses (e.g. infiltration from or drainage to the rivers) - **Contaminant Transport** - if contaminant is transported by advection (defined by characteristics of the contaminant - see 2.2) preliminary assumption on direction of contaminant transport can be made.

Table 6.2 continued

Results of SC given as fulfilment of SC objectives	SC Output	SC Output form - MC Input sources	Relevance for Model conceptualisation item (for development of mathematical model)
1.2.4 Surface water	Principal output: - distribution on surface water courses over the area, measured discharges and levels - estimated interaction with GWS (infiltration/drainage) Additional output: - riverbed levels, wetted perimeters, drainage classes, etc	- **hydrology II map** (if not too many data, than could be combined with the **hydrology I** map); includes sea, lakes, rivers, brooks, springs, drainage channels, etc. - numerical output in the **budget table** - (if not steady state) **hydrograph** and table showing temporal (mind the time step!) distribution (levels and discharges) - table with **river parameters** (riverbed levels, wetted perimeters, drainage classes, resistivity)	- **Model Boundaries** - large bodies of surface water could play a role of a physical boundary in the model. - **Boundary Conditions** - in the case described above (surface water - physical boundary), surface water courses parameters (level, riverbed resistivity) influence boundary conditions. - **Sinks and Sources** - surface water courses parameters (level, riverbed resistivity) influence system inflow/outflow at upper system boundary.
1.2.5 GW withdrawal 1.2.6 Artificial recharge	distribution on pumping/recharge wells (galleries) over the area, measured quantities	- **GWLhead** map with distribution of **pumping/recharge wells** (galleries) for each aquifer where it takes place. - (if not steady state) hydrograph and table showing temporal (mind the time step!) distribution	- **Sinks and Sources** - withdrawal\recharge wells influence system inflow/outflow at upper system boundary.
1.2.7 Groundwater Inflow/Outflow	estimated GW Inflow/Outflow	- estimated GW **inflow/outflow** in the budget table Additional output form: - **budget** map with marked hydroisohypses that were used for calculation	- **Boundary Conditions** - at the GWS boundaries which do not resemble aquifer boundaries, the groundwater inflow and/or outflow have to be estimated. These figures are needed as first estimation of the model Boundary Conditions.

Table 6.2 continued

Results of SC given as fulfilment of SC objectives	SC Output	SC Output form - MC Input sources	Relevance for Model conceptualisation item (for development of mathematical model)
1.3. System internal conditions	Information on characteristics of GWS that determine mass transport and (partially) mass transfer processes in the system	See the individual items	- **Model parameters** based on physical properties on the GWS
1.3.1 Conductivity/ Resistivity	hydraulic conductivity of aquifers and resistivity of aquitards and riverbeds	- **conductivity** map for each aquifer - **resistivity** map for each aquitard - **resistivity** map for riverbeds	- **Model parameter 'k'** - main mass transport parameter, controls advection - seepage velocity (with porosity) and dispersion (with dispersion and diffusion coefficients) in the aquifer. - **Model parameter 'c'** - controls advection in the aquitard. - **Model parameter 'c'** - controls advection in the riverbed.
1.3.2 Porosity	porosity of the hydrogeological units	- table with overview of (total and effective) porosity values per hydrogeological unit.	- **Model parameter 'n$_t$'**, total porosity (total void space in the porous medium) required for math. representation of retardation (due to sorption). - **Model parameter 'n$_e$'** effective porosity (interconnected portion of the void space in the porous medium) required for math. representation of advection - seepage velocity.
1.3.3 Storativity	storage capacity of hydrogeological units	- table with overview of storage coefficients (for confined and semi-confined units) and specific yield for unconfined unit(s).	- **Model parameter 's$_s$'** storage coefficient (required for transient calculations). - **Model parameter 's$_y$'** specific yield (required for transient calculations).
1.3.4 Diffusivity / Dispersivity	information of mass (fluid) spreading and mixing due to concentration gradient and unresolved differences in permeability distribution	- table with overview of values for effective diffusivity and longitudinal and transversal **dispersivity** coefficients per hydrogeological unit.	- **Model parameters** required math. representation of mass transfer (diffusion and dispersion) processes: D - effective diffusion coefficient α_l - longitudinal dispersion α_t - transversal dispersion (see comments)
1.3.5 Organic carbon (oc) content/particle mass density (pmd)	information on oc content in hydrogeological units and dry bulk density	- table with overview of values of oc content given as fraction of the total sediment (**foc**) per hydrogeological unit & pmd values.	- **Model parameter 'R'** 'foc' fraction organic content and 'dbd' dry bulk density are required for calculation of retardation due to sorption. (see comments)

Table 6.2 continued

Results of SC given as fulfilment of SC objectives	SC Output	SC Output form - MC Input sources	Relevance for Model conceptualisation item (for development of mathematical model)
2.SC POLLUTION ESTIMATION	Information on contaminant (its characteristics and presence in the GWS) and GW quality	See the individual items	- **Model parameters** based on (chemical) properties of the contaminants - **Contaminant transport** - **Background information** (Groundwater quality, Contaminant identity, ...)
2.1. Basic GW characteristics 2.1.1 Hardness 2.1.2 pH 2.1.3 EC 2.1.4 TDS	Basic Information on groundwater quality	- **Groundwater quality** table or maps (depending on data availability- quantity and variability)	- **Groundwater quality** (Background information) Additional use: - **Flow Pattern** and **Flow Regime** (see Anderson p.38 - needs working out) - **Sinks and Sources** - interaction between GWS and surface water courses (see Anderson p.38 - needs working out)
2.2. Pollution characterisation	Information on pollutant characteristics that (partially) determine mass transfer processes in the system	See the individual items	- **Model parameters** - **Background information**
2.2.1 Chemicals	information on indicator parameters (chemical) selected for pollution monitoring (selection is based on prevalence, mobility and toxicity)	- **Contaminant Identity** table showing overview of indicator parameters, max. concentration observed, toxicity indicators, regulatory levels, mobility indicator.	- **Contaminant identity** (Background information) required for selection of contaminant for transport modelling (modelling state variable)
2.2.2 Concentration	information on concentration of indicator parameters	- table and/or **concentration** map(s) (depending on data availability- quantity and variability) - (if not steady state) diagram and table showing temporal conc. distribution	- **Contaminant identity** (see above) - **Contaminant transport** - initial loading of contaminant concentration (modelling state variable)
2.2.3 Toxicity	information on toxicity of contaminants (possibly) present at the site	- table showing toxicity indicators and regulatory levels (required for **Contaminant Identity** table)	- **Contaminant identity** (see above)

Table 6.2 continued

Results of SC given as fulfilment of SC objectives	SC Output	SC Output form - MC Input sources	Relevance for Model conceptualisation item (for development of mathematical model)
2.2.4 Mobility	information on mobility of contaminants (possibly) present at the site	- **table** showing contaminants mobility (if solubility is used as indicator of mobility than the table identical to Solubility table) (required for **Contaminant Identity** table)	- **Contaminant identity** (see above)
2.2.5 Solubility	information on solubility of contaminants (possibly) present at the site	- **Solubility** table	- **Model parameter**- can be used for K_{oc} estimation (optional 2.2.6) - [retardation (comments)] - **Contaminant transport** - initial loading of contaminant concentration (pessimistic upper limit)
2.2.6 K_{ow}	information on octanol/water partition coefficient	- **K_{ow}** table	- **Model parameter** - can be used for K_{oc} estimation (optional 2.2.5) [retardation (comments)]
2.2.7 Half-life	information on rate of decrease of contaminants due to degradation	- **half-life** table	- **Model parameter** - first-order degradation rate constant
2.3 Pollution delineation 2.3.1 Source delineation 2.3.2 Plume delineation	information on distribution of contaminant selected for modelling	- **source delineation** map [detailed, showing contaminant concentration distribution-observed and interpolated (regionalised) values] - **plume delineation** map (the GWS representation scale, showing: ditto above)	- **Contaminant transport** - estimation of initial loading of contaminant concentration

Table 6.2 continued

Results of SC given as fulfilment of SC objectives	SC Output	SC Output form - MC Input sources	Relevance for Model conceptualisation item (for development of mathematical model)
3. ADDITIONAL CHARACTERISATION	various pieces of information that are (or can be) useful for the modelling	See the individual items	See the individual items
3.1. Budget calculation	outcome of budget calculation	- numerical output in the **budget** table	- **Boundary conditions & sink and sources** (adjustment & estimation of model variables initial values; (later: model calibration);
3.2. Management practice	information on location and characteristics of drainage an/or treatment facilities	- drawn on the **source delineation** map and described	- **Contaminant transport** - estimation of initial loading of contaminant concentration - **Background information**
3.3 Agriculture practice, nature resorts, settlements	Potential pollution targets (next to well fields)	- locations or distributions drawn on **GWS overview** map	- **Model boundaries**
3.4...			

Comments:

- content of SC is not definitive; among the other factors, it depends on processes included;
- hydrographs, charts and tables that show temporal distribution of the variables (gw head, pumping rates, etc.) are always desirable while making a decision on: flow in the model (steady-state or transient), time-step (transient modelling) or selection of time period (steady-state modelling). They are, however, compulsory if the decision to perform transient modelling is taken;
- there is a number of tasks that can be treated as 'Additional (site) characterisation' tasks - for modelling purposes' or 'Model Conceptualisation' tasks (e.g. budget calculations, determination of retardation factor, selection of dispersivity coefficient, etc). (In the table, the 'budget calculation' is regarded as SC task because it is preferable to be carry it out regardless of modelling.) Nevertheless, in the DSS most of these tasks are (or will be) independent modules that can be invoked either from the SCM, GMM or an other DSS KBM.

Table 6.2 shows when (where) individual SC results can be required during the definition of MC items; for example groundwater level/head is needed for definition of Flow Pattern, Boundaries, Boundary Conditions, etc. (see the table). The modeller, however, approaches to the problem of MC from the side of MC items, meaning that all individual SC results that might contribute to definition of an MC item should be systematised per item. The example of MC item (Model Boundaries) and required pieces of information (i.e. individual SC results) are given in the previous paragraph. The systematisation per MC item is provided in the GMM and will be presented in Section 6.4.

The second MC task (decisions on Model Complexity directed by MO) also finds a place within the same framework as given in Figure 6.2. It should be worked out which MC items are subjected to further conceptualisation and in which way. For example:

- presence of bedding planes and laminae within a sequence of sediment layers can cause anisotropy of groundwater flow. This should be considered in the light of Model Complexity (posed by MO) during the conceptualisation of the Model Matrix, and consequently during the conceptualisation of Model Parameters (see the figure).

The outlined MC content is compared with those suggested by other authors in Table 6.3. Common for all reviewed concepts is that they are given in a very general form; an exception is the concept given by Anderson and Woessner (1992), but only to a certain extent (too little attention is paid to the MC input). The reviewed concepts bring few novelties so no further discussion is found necessary.

6.3.3   Modelcode Selection (MSEL)

Selection of a modelcode nowadays receives appropriate attention (see software overview, Section 2.2).Therefore, no special attention was paid to Modelcode Selection (MSEL) at this stage of the GMM development. That decision was also based on the fact that the expert systems (or databases) for selection of the model code can be used independently, hence without integration into a DSS. Additionally, selection of modelcode in practice is, for the majority of problems, primarily a question of availability and personal preferences.

Nevertheless, the main issues related to MSEL will be addressed here in order to point out a direction of further GMM development (related to MSEL), and 'preserve a link' between the steps of the Modelling Protocol (MC and MD).

Selection of modelcode begins by setting up the *mathematical model*; subsequently the most suitable *solution technique* should be chosen. The selected modelling technique is in most cases encoded in several modelcodes; which among them should be selected depends on a number of factors. In the GMM, these factors are divided into three groups:

Table 6.3 Various views on model conceptualisation

Model Conceptualisation items (GMM)	Anderson & Woessner (1992)	NRC (1990)	Bear & Verujit (1987)	OSWER (1994)	ASTM (1993a)	Bear et al (1992)
Model geometry	- defining hydro-stratigraphic units (also Model Parameters)	- the size and shape of the region of interest	- identify the materials contained in the flow domain - identify the boundaries of the modelling domain - make assumption concerning the homogeneity and anisotropy (also 3)	- hydrostratigraphy - hydrologic boundaries	- geological framework	- geometry of the boundaries - the kind of solid matrix - presence of sharp fluid-fluid boundaries
Model variables	- preparing a water budget - defining the flow system	- the boundary and initial conditions for the region	- the behaviour of external domain and interaction at common boundary between the domains - identify sink and sources	- Boundary conditions - ground-water flow system - water sources and sinks - Contaminant source, loading and areal extent	- hydrologic framework (also 1) - sources and sinks	- the relevant state variables - sources and sinks of water and relevant contaminants - initial conditions (Also model Param.) - boundary conditions
Model parameters		- the physical and chemical properties that describe and control the processes in the system		contaminant transport and transformation processes [also 4 and Define Purpose]	- hydraulic properties	
Modelling background info				- background chemical quality - Contaminant identity		- the properties of the water (homogeneity, effect of dissolved solids and/or temperature on density and viscosity [also Define Purpose]
Other	- only flow and advective transport considered		- assume the type of flow regime [Define Purpose]; only flow and advective transport considered			- model dimensions [Define Purpose] - the flow regime - [Define Purpose]

1. The first group is related to Modelling Objectives (MO), expressed through *Model Complexity* (MCO), which is established during DP and MC;
2. *Modelcode reliability* is shown separately to emphasise the importance of this factor (usually underestimated in practice);
3. *Modelcode availability* is also shown separately to emphasise the importance of this factor.

Mathematical Model

'The mathematical model contains the same information as the conceptual one, but expressed as a set of equations which are amenable to analytical and numerical solution' (Bear et al, 1992). The same authors gave 'the complete statement of a mathematical model', consisting of the following items:

- A definition of the geometry of the considered domain and its boundaries;
- An equation (or equations) that express(es) the balance of the extensive (i.e. field) quantity (or quantities);
- Flux equations that relate flux(es) of the considered extensive quantity(ies) to the relevant state variables of the problem;
- Constitutive equations that define the behaviour of the fluids and solids involved;
- An equation (or equations) that expresses initial conditions that describe the known state of the considered system at some initial conditions, and
- An equation (or equations) that defines boundary conditions, i.e. the interaction of the considered domain with its environment.

If the content of the mathematical model is compared with the outcome of the conceptual model (Figure 6.2), a clear link can be seen between these two modelling steps, established through geometry, parameters and variables. The modeller should select a set of equations that describes the problem at hand in the most appropriate way and use that information as a starting point in the MSEL. Once the mathematical model is set up, a solution technique has to be selected.

Solution Technique

The very first question is whether analytical or numerical solution is more suitable. The selection depends mostly on modelling objectives and model complexity. Anderson and Woessner (1992) stated that selection also depends on the 'conceptual views of the groundwater system'; 'aquifer viewpoint' is especially suited for well hydraulics and is the basis for many analytical solutions, whereas the 'flow system' viewpoint asks for a numerical approximation. In practice, analytical models are less used than numerical. However, their application is always appreciated while verifying the modelcode and the model results.

Special attention should be paid to the approximations and assumptions of numerical techniques and their possible impact on the modelling objectives and credible representation of site-specific conditions (for further discussion see Section 2.3.2).

Model Complexity Outcome

Model Complexity (MCO) is based on Modelling Objectives (MO). It is defined gradually in several modelling steps (DP, MC, MD and MCAL). In principle, MCO is crucial factor in modelcode selection, although selection of modelcode could sometimes induce revision of demanded model complexity. From the analysis of software that has been developed for modelcode selection (KGM and Model Expert) it would appear that (rather general) information required for modelcode selection could already be obtained during the DP. (It should, however, be taken in consideration that MC could lead to revision of MO and consequently MCO.)

In the Model Expert, modelcode selection is based on following issues:

- Model availability: public domain or proprietary codes,
- Type of medium: fractured rock, porous media only, or both,
- Simulation type: flow, transport, flow and transport, or soil/aquifer parameter estimators,
- Soil materials: homogeneous soils or heterogeneous systems,
- Soil saturation: saturated, unsaturated conditions or both,
- Aquifer type: confined aquifers, unconfined aquifers or both,
- Fluid phases: multiphase, single phase or both,
- Dimensions: one, two and three dimensional models,
- Chemistry: organics, inorganics and radioactive substances,
- Verification: extensive, moderate and limited, and
- Model cost and number of users.

A substantial part of KGM content is used in development of the Model Complexity Module (MCM) and will be discussed in Section 6.4.

Modelcode reliability

In order to assess modelcode reliability OSWER recommends a review of:

- Peer reviews of the modelcode (e.g. a formal review process by an individual or organisation acknowledged for their expertise in groundwater computer models);
- Verification studies (e.g. evaluation of the model results against laboratory tests, analytical solutions, or other well accepted models - see also Footnote 6 on page 16);
- Relevant field tests (i.e. the application and evaluation of the model to site-specific conditions for which extensive data sets are available);
- The modelcode acceptability in the user community as indicated by the quantity and type of use.

Availability

Very often the availability of the modelcode plays an important role in modelcode selection. The modelcode that is already available will be often used, without comparing it with other existing codes. Moreover, public domain modelcodes are much more in use than commercial modelcodes, even in the cases where the latter are evidently more suitable for the problem at hand.

According to OSWER, the modeller should also take into account the availability of the model binary code, the model source code, pre- and postprocessors, existing data resources, standardised data formats, complete user instruction manuals, sample problems, necessary hardware, transportability across platforms, user support and key assumptions.

Last, the trade-off should be assessed and implemented between modelcode performance (e.g. accuracy and processing speed) and the human and computer resources required to perform the modelling.

6.3.4   Model Design (MD)

The principal inputs are MC output and the (selected) Modelcode; the MCO summary (DP outcome updated after conducting MC and MSEL) should be consulted as well. MD output is a 'ready-to-run' model that should be calibrated and tested (sensitivity analysis); then the prediction runs can be carried out. The Model Design step is also called 'Model Construction' (ASTM, 1993a, 1995a) or 'Model Setup and Input Estimation' (OSWER, 1994).

According to ASTM (1995a) the groundwater flow Model Construction is 'the process of transforming the conceptual model into a mathematical form ' (where ' groundwater flow model 'typically consists of the data set and computer code'). Further, the model construction process 'includes building the data set utilised by the computer code'. Finally, 'fundamental components of the groundwater flow model include dimensionality, discretisation, boundary and initial conditions, and hydraulic properties'. (In the Guide for documenting groundwater flow application (ASTM, 1995b) the following items (related to Model Construction) are addressed: model domain, hydraulic parameters, sources and sinks, boundary conditions, selection of calibration targets and goals and numerical parameters.)

OSWER does not give the precise content of 'Model Setup and Input Estimation'. However, a number of items are listed (in the same document) as needing to be documented in the modelling report, covering 'Groundwater model construction'. Those items are: code modification, geologic representation, flow representation, data averaging procedure, input estimation procedure, model grid, hydraulic parameters, chemical parameters, boundary conditions, water budget and simplifying assumptions. ('Uncertainty Analysis' is given as a separate reporting item that follows Model Construction and precedes Calibration).

Anderson and Woessner (1992) consider Model Design as an independent step of the Modelling Protocol. Description of MD is, however, spread over several chapters; grid design is given as a follow up of the Model Conceptualisation (in the same chapter), whereas 'boundaries', 'sources and sinks' and ' special needs for transient simulation' form separate chapters.

The brief overview that is given above shows a lack of consensus (and systematics) while describing MD. MD could be seen (as OSWER suggested) as a modelling step consisting of two basic tasks: the model setup and input estimation; for example, a grid has to be constructed and data have to be (estimated for, and) assigned to each grid cells. Although the latter in principle follows the grid construction (i.e. the model setup), the task are not completely sequential because the grid is constructed taking into consideration the required input estimation. Accordingly, the MD should rather be considered as *an unique task that includes determination of model items (or components)* such as grid, boundaries etc. Determination of each item involves use of MC outcome, modelcode and (probable estimation and) allocation of model input. Therefore, each item of MD should, beside the description (of the design task), also contain a list of related inputs and link(s) with the modelcode. [9, 10] All possible interrelated items should be mentioned (and linked) within the task description and also listed separately (as a link list).

The MC outcome provides input for the MD, therefore *model geometry*, *model variables* and *model parameters* could be considered as main MD components, where information contained in MC 'modelling additional info', can be grouped (if needed?) in *additional specification*. Model geometry, model variables and model parameters will be elaborated in following three sections.

Model Geometry

'In the numerical model the continuous problem domain is replaced by a discretized domain consisting of nodes and associated finite element blocks (cells) of finite elements' (Anderson & Woessner, 1992). In the other words, the model geometry (matrix and model boundaries) in the conceptual model are represented by the nodal grid, i.e. 'the nodal grid forms a framework of the numerical model'. The way of representing GWS layers (defined in the conceptual model) by a grid in the numerical model depends on model complexity, grid type (finite elements, finite difference) and other modelcode specifics, type of numerical model (2-D, quasi 3-D, 3-D), etc. The modeller should perform two basic tasks: lay out the *grid* and define *model layers* (number, type(s) and thickness).

---

[9] Some of 'related inputs' are shown as <u>underlined text</u> in the following sections as an illustration of their function in the GMM; in the GMM these terms are virtual links with the MC outcome (maps, tables, diagrams).

[10] As Model Design reflects the characteristics of the selected modelcode (e.g. FE or FE grid) some information on MODFLOW (related to Model Design) and the links with PMWIN interface are contained in GMM.

If the modelcode prescribes an unique *grid* for all the layers (MODFLOW does, like all the other modelcodes with a finite difference grid), the grid has to be specified only once. The first step is to lay out the grid boundaries that should correspond (as much as possible) with the area (to be modelled) specified during the Model Conceptualisation, in order to reduce the number of inactive cells. Problems with inactive cell do not appear if the grid is constructed of finite elements (specifics of the selected grid type should therefore be analysed). The conditions at the boundaries and their representation in the numerical model have also to be taken in account while setting grid boundaries. (Possible representation of boundary conditions should have already been taken into consideration during the MC.) At the same time, the right decision has to be taken on grid orientation. Useful recommendations for grid orientation are given in Anderson and Woessner (1992).[11] Firstly, the grid should be laid over the GWS overview map that contains basic data such as extent of the investigated area, location of (potential) pollution source(s), (potential) pollution recipient(s), etc. The grid axis should coincide (as much as possible) with principal direction(s) of the hydraulic conductivity tensor, therefore conductivity maps and profiles need to be consulted.

The next step is selection of grid density. 'Selecting the size of the nodal spacing is a critical step in grid design' (Anderson & Woessner, 1992). The grid density should be a compromise between numerical accuracy and practicality (complexity and data availability). As the grid density is the function of variability of the groundwater table or potentiometric surface, Ground Water Level/head (GWL/head) map(s) should be consulted. Variability in areal recharge, pumping, and recharge from (or discharge to) rivers should also be taken in consideration while specifying the grid density. Required information can be found on hydrology map(s), GWL/head map(s) and related tables and diagrams. Trade-off between the grid density (a cell size) and time steps (time parameters) is needed when a transient model is designed.

The number of hydrogeological *layers* is primarily defined by the conceptual model, developed during the Model Conceptualisation (MC). Results of MC are given on the maps showing thicknesses of hydrogeological units and hydrogeological cross-sections. In the conceptual model of the GWS each lithostratigraphic unit (that shows a significant extent with respect to a considered scale) is usually represented by a separate hydrogeological unit. Some changes, however, may occur during the conceptualisation for the purpose of mathematical modelling. If there are, for instance, significant vertical head gradients, two or more modelling layers should be used to represent a single lithostratigraphic unit. On the contrary, some discontinuous or just locally developed units (i.e. lenses) can be disregarded. Finally, neighbouring stratigraphic units with similar lithology can be combined into a unique layer.

The number of modelling layers also depends on the type of model that has been selected for

---

[11] The section on grid orientation, like several other sections from the book of Anderson and Woessner (1992), is scanned and included in the GMM. Selected sections are concise enough and information rich, so no modification in terms of shortening or rewriting has been found necessary.

numerical presentation of the GWS; for example, a confining unit will be represented as a separate layer in the fully 3-D model, which is usually not the case if the quasi 3-D model is selected.

In PMWIN, the layers can be designed as always confined, always unconfined or convertible (capable of being either confined or unconfined). Layer types are explained in detail in PMWIN on-line help. Analysis of the <u>GWL/head maps</u> is required before assigning the layer types. <u>GWL/head hydrographs</u> should also be analysed in cases of possible layer convertibility (from confined to unconfined and vice versa).

The thickness of a model layer is the third (vertical) grid dimension, and can be defined in PMWIN in two (indirect) ways; the user can specify elevations of the <u>bottom and top of the layer</u>, or specify transmissivities which are equal to the hydraulic conductivity of the layer multiplied by the layer thickness.

Model variables

The most important model variables are head, flux and concentration (flux can be also seen as the derivative of a dependent variable). Model variables define inputs and outputs from the model at external and internal boundaries (*input/output variables*) and the state of the mass transport and transfer in the model (*state variables*).

Specification of *boundary conditions* means in practice specification of conditions at external boundaries. Conditions at the external boundaries are represented in the model by specified head boundaries, specified flow boundaries (a no-flow boundary is set by specifying flux to zero) and/or head-dependent flow boundaries. All the option listed can be used to simulate sources or sinks (or internal boundaries) in the interior of the problem domain, meaning that not only *recharge/discharge wells*, but also *rivers*, *lakes* and *drains* are considered as 'sources and sinks'. In addition, groundwater *recharge* to the water table is also a source term in areal two-dimensional models and in profile and three-dimensional models that use the Dupuit assumptions to simulate flow in the upper layer of the model. (*Evapotranspiration* can, accordingly, be regarded as a sink term.) However, in some profile models and in fully 3-D models groundwater recharge to the water table is a boundary condition, and groundwater table (otherwise a state variable) forms a part of the boundary.

Boundary conditions. The main tasks related to representation of boundary conditions in numerical model are:

− to select the type of mathematical representation (i.e. specified head, specified flow or head-dependent flow),
− to assign the values for flow boundaries.

Representation of boundary conditions in a numerical model is a Model Design task, although it is (should be) already taken in consideration during the Model Conceptualisation; boundaries and boundary conditions in the conceptual model are defined bearing in mind their possible representation in the numerical model.[12] Accordingly, the major part of information required for selection of boundary conditions in the model is available from the Model Conceptualisation results. The MC results contain description of boundaries and boundary conditions. Boundaries are also represented graphically on the <u>GWS overview map</u>, <u>hydrology map</u> (e.g. surface watercourses) and <u>GWL/head map</u> (e.g. hydroisohypses). Some information can also be found in the <u>budget table</u> (e.g. flux calculation). If transient modelling is performed, the GWL and river level <u>hydrographs</u> should also be consulted.

Recharge/discharge wells. The modeller has to specify locations of the wells from the <u>GWL\head maps</u> for aquifers where injection/pumping takes place. These maps should contain well locations. In quasi 3-D and 2-D areal models, the node represents the thickness of the aquifer. Hence it is implicitly assumed that the well penetrates the full thickness of the aquifer. MODFLOW can simulate wells that penetrate more than one model layer. In this case, the injection or pumping rate for each layer has to be specified. Another possibility to simulate multi-layer wells in MODFLOW is to set a very large vertical hydraulic conductivity or vertical leakance (for details, see MODFLOW on-line help).

The pumping/injection rates (for a selected time period) can be found in the <u>overview table</u> produced as a result of Model Conceptualisation. If a transient model is developed, the variable rates can be assigned to each stress period (see Time parameters). In that case, the <u>diagram</u> showing <u>temporal distribution</u> of <u>pumping/injection</u> rates should be consulted .

Surface watercourses and drains. Recharge from (or discharge to) the rivers or reservoirs is usually handled as head-dependent flow. The main difference between rivers and drains is that the latter do not allow recharge to the aquifer (only discharge to the drain is possible). Information required for modelling of rivers and drains can be found on the <u>hydrology map</u>, a <u>table</u> containing river/drain parameters (riverbed levels, wetted parameters, resistivity, etc) and <u>hydrographs</u> showing temporal distribution of river levels. If the rivers play a special role in the model, a map can prepared showing riverbed resistivities.

Recharge in the numerical model refers to the flux across the water table. The flux can be treated as internal source or boundary condition. In both cases it has to be entered into the model. Values of recharge should be obtained from the <u>hydrology map</u>, or simply from the <u>budget table</u>, if recharge is assumed to be uniform. If precipitation varies substantially in time (transient

---

[12]In some cases representation of boundary conditions in the numerical model is practically already defined during the MC; impermeable fault zones and groundwater divides, for example, will certainly be represented by no-flow boundaries. Therefore the section 'Setting Boundaries' (Anderson & Woessner, 1992) deals also with the representation of boundary conditions in the numerical model. More about specifics of various boundary condition types in the numerical model is given in the section 'Simulating Boundaries'.

modelling), a diagram showing the temporal distribution of precipitation should be analysed, and data from the adjacent table utilised.

Evapotranspiration in MODFLOW is treated as one of head-dependent fluxes (from the system). Information required for the model is obtained from the same sources as precipitation (hydrology map, budget table and diagram showing temporal distribution of evapotranspiration (if needed)).

The *state variables* show 'state' of the groundwater system and reflect changes caused by input/output variables (hydrological stresses). Basic GWS state variables are:

– groundwater level/head and
– contaminant concentration

The groundwater model calculates state variables and their (spatial and/or temporal) distribution which are, in most cases, the main outcomes of the modelling process. During the Model Design the initial values of state variables should be entered into the model. Unlike some other modelcodes, MODFLOW asks for initial values of groundwater level/head even if only state-state calculations are to be performed ('starting values'). A level/head calculated in a steady-state calculation is used as the initial level/head for transient calculations. In the case of drawdown simulations, static steady state conditions can be assumed, where the head is constant throughout the system and no flow of water takes place. In most cases, however, the modeller should enter levels/heads variable in space, that represent dynamic average steady-state conditions.

Information on groundwater level/head can be found on GWL/head maps. These maps are usually produced by (semi)automatic interpolation of point observations; which do not always provide the most realistic spatial distribution of the levels/heads. If some adjustments are needed, they should be carried out before data are transferred to the model grid. Since the level/head has to be assigned to each cell of the model, transferring usually means additional interpolation (this time, automatically done). Temporal distribution of the levels/heads (required for transient calculations) can be found on hydrographs and adjacent tables.

Initial contaminant concentration can be found on concentration maps and the contaminant identity table. Temporal concentration distribution is shown on the diagram and adjacent table.

Model Parameters

Hydraulic conductivity and resistivity. Hydraulic conductivity has to be assigned to all aquifers defined in the model. In MODFLOW, transmissivity of the aquifers can be assigned directly by the user or calculated by the PMWIN. Values of hydraulic conductivity are available on conductivity maps. Instead of resistivity values for the aquitards, MODFLOW asks for a leakage term (1/resistivity). Leakage can be calculated by the PMWIN using vertical conductivities and thicknesses of the layers, or entered by the user. In the case of (quasi) 3-D models, leakage has to be entered by the user.

Storativity is required for transient calculations in a form of a storage coefficient (confined layers) or specific yield (unconfined layers). In PMWIN a storage coefficient can be specified by the user or calculated by PMWIN (using user specified storage and layer thicknesses). Values of storativity (or calculated storage coefficient and/or specific yield) can be obtained from the overview table.

Effective porosity is required for calculation of average velocity in particle tracking and transport modelling. Values of porosity can be obtained from the overview table.

If *diffusivity* and *dispersivity* (mass transport process) are found important enough (Model Complexity) to be included in the numerical model, the parameters that represent these processes have to be assigned. Parameters are defined by using the DE1 module and presented in the overview table.

If *retardation* due to sorption is found important enough (Model Complexity) to be included in the numerical model, the retardation factor has to be estimated. The retardation factor can be estimated by using the RE1 module. However, PMWIN also calculates the retardation factor, so RE1 can be used to define K_d, the linear sorption coefficient required by PMWIN.

If *degradation* (due to first order irreversible rate reactions) is found important enough (Model Complexity) to be included in the numerical model, the first-order rate constants have to assigned. Constants (given in the form of half-life times) are presented in the overview table.

Time parameters are required for transient modelling. Stress periods, time steps and transport time steps have to be assigned to MODFLOW. The section on transient simulation (Anderson and Woessner) and ASTM documents (1994a, 1994b) are scanned and included in the GMM.

6.3.5 Calibration and model predictions

Calibration is a process of refining the numerical model (i.e. representation of hydrogeological framework, hydraulic properties and boundary conditions in the numerical model) to achieve a desired degree of correspondence between the model simulations and observations of the groundwater flow system. In other words, it is a procedure for finding a set of parameters and hydrological stresses that produces simulated heads and fluxes that match field-measured values within an acceptable range of error. The main calibration steps are:

- establishment of calibration targets,
- identification of calibration parameters and

- history matching.

Establishment of calibration targets. According to ASTM (1996) this step consists of establishment of calibration targets and associated acceptable residuals or residual statistics. The calibration target itself consists of the best estimate of a value of groundwater head or flow rate. Slightly different terminology is used by Anderson and Woessner whose calibration value with its associated error form the calibration target. Accordingly, calibration values should be selected from field data, and than the error in selected values should be estimated in order to define calibration targets.[13] *Calibration values* or *sample information* are field measured values of heads and fluxes, that might contain various types of errors (due to interpolation, scale, transient effects, etc). In order to establish reliable estimation of measurement error, all pieces of information relevant for calibration values (e.g. bottom elevation of dry wells, estimates of baseflow, etc.) should be identified and their relation with the measurement error estimated.

Acceptable error of calibration target depends on the measurement error and on the accuracy (precision) required. Due to the many approximations employed in modelling and the errors associated therewith, it is usually impossible to make a model reproduce all head measurements within the error of measurements. Therefore, the range of acceptable calibration target error must be larger than the range of the measurement error. In general, however, the acceptable error should be a small fraction of the difference between the highest and lowest heads across the site. Some additional advice and recommendations related to establishment of acceptable error (such as use of kriging and distinct hydrological conditions) are given in the scanned ASTM document (1993b).

Identification of calibration parameters. Calibration parameters are groups of hydraulic properties or boundary conditions whose values are adjusted as a group during the calibration process. Calibration parameters can be actual inputs to the numerical model or quantities used in the preprocessing phase, i.e. input preparation (e.g. the thickness of streambed silt deposits as used to calculate the leakance of river nodes). Prior information on calibration parameters is usually scarce and influenced by uncertainty. The coefficient of variation (standard deviation divided by expected value) may be used to quantify the uncertainty associated with each piece of prior information. The plausible ranges (ranges of possible realistic values) must be established for each calibration parameter before beginning any simulation.

History matching. History matching is accomplished by using the trial and error method to achieve a rough correspondence between the simulation and the physical hydrogeological system, and then using either the trial-and-error method or an automated method to achieve a closer correspondence. (Basic information on these methods, as given in ASTM guide (1996) is

---

[13] Anderson and Woessner (1992) provided a summary of the procedures to be used 'before and after' calibration (in that approach, the calibration is probably considered as just a pure running of the model). In a such framework, calibration target estimation and parameter identification are carried out 'before' calibration.

available in the GMM.) The calibration parameters are varied within the specified range until the model produces a satisfactory level of correspondence between the simulation and the physical hydrogeological system. The 'calibration sensitivity analysis' should be employed, being a systematic way to find values which result in the lowest residuals (errors).

Qualitative and quantitative techniques have been developed for comparison of the simulation results with the site-specific information. Qualitative techniques are related to:

- general flow features (assessing the correspondence between the overall patterns and modelled head contours);
- input parameters (assessing whether the model input parameters fall within the previously established ranges of reasonable values), and
- hydrological conditions (evaluating the number of distinct hydrological conditions that a model is capable of reproducing). This is also called 'model verification' and can be performed if a second independent (transient or steady-state) data set is available. More detail on this qualitative technique is given in the (scanned) section Model verification (Anderson and Woessner, 1992).

Quantitative techniques include:

- calculating piezometric head residuals,
- assessing correlation among head residuals, and
- calculating flow residuals

The standard guide for 'Comparing Ground-Water Flow Model Simulations to Site-Specific Information' (ASTM, 1993a) contains an extensive description of qualitative and qualitative techniques used in history matching. The guide is, together with associated examples, included in the GMM. The GMM also contains a detailed description of sensitivity analysis of a calibrated model. Purpose and content of sensitivity analysis are briefly outlined below.

Sensitivity Analysis

'The purpose of a sensitivity analysis is to quantify the uncertainty in the calibrated model caused by uncertainty in the estimates of aquifer parameters, stresses and boundary conditions' (Anderson and Woessner, 1992). According to ASTM (1994c), the sensitivity analysis is carried out after a model has been calibrated, and after the model is used to draw conclusions about a physical hydrogeological system. The purpose of sensitivity analysis is, therefore, to identify which model inputs have the most impact on the degree of calibration and on the conclusions of the modelling analysis. Accordingly, if variations in some model input results in insignificant changes to the degree of calibration, but cause significantly different conclusions, then the mere fact of having used calibration does not mean that the conclusions of the modelling study are valid.

The main Sensitivity analysis steps (ASTM, 1994c) are:

- identification of inputs to be varied,
- execution of simulation,
- graphing results, and
- determination of the type of sensitivity.

Calibrated values for hydraulic conductivity, recharge, storage parameters and boundary conditions are usually analysed by changing one parameter value at a time. However, two or more parameters can be analysed simultaneously in order to determine the widest range of plausible solutions. Describing sensitivity analysis, Anderson and Woessner (1992) also refer to various techniques (including stochastic modelling) performed when using automated calibration codes. According to ASTM (1993b) these are a part of 'calibration sensitivity analysis' and not 'sensitivity analysis'; the latter includes the effects of varying inputs on model predictions as well as on the calibration and therefore provides a method of distinguishing between significant and insignificant degree of sensitivity.[14] That also means that ASTM does not recognise 'prediction sensitivity analysis' (see next section).

Model Predictions

Owing to uncertainty in future hydrological stresses, model predictions introduce new (additional) errors in simulation. When a physical system is subject to new stresses (as during the application of a remediation strategy), errors in the conceptual model which had little impact during the calibration phase may become dominant sources of error for the prediction phase. Therefore it is important to explore the implications of uncertainties in model input parameters and conceptual assumptions on model prediction. That is done by testing the effect of uncertainty in the calibrated parameters, i.e. by performing sensitivity analysis (as described in the previous section) for at least one of the predictive simulations.

One of the difficult tasks in predictive modelling is selection of the length of time for which the model should produce accurate prediction. It has been suggested that a predictive simulation should not be extended into the future for more than twice the period for which calibration data are available. Selection of simulation period and other issues related to predictive modelling will be thoroughly investigated in the next stage of GMM development.

---

[14] Anderson and Woessner (1992) and OSWER (1994) use the term 'calibration sensitivity analysis' for assessing the effect of uncertainty on the calibrated model and the term 'prediction sensitivity analysis' for assessing the effect of uncertainty on the prediction. ASTM combines these two concepts in 'sensitivity analysis' because 'only by simultaneously evaluating the effects on the model calibration and predictions can any particular level of sensitivity be considered significant or insignificant'. (the term 'calibration sensitivity analysis' is used by ASTM for a systematic procedure for finding the best value set (that produces the lowest residuals) during the calibration (see History matching).

6.3.6 Modelling Report

The modelling report includes written and graphical presentations of model objectives and assumptions, the conceptual model, code description, model design (construction), model calibration, predictive simulations and conclusions (ASTM, 1995b).

1. Introduction

  1.1 General setting of the site and the problem description (the main part of the required information should be available from the SC summary)

    General setting description; very briefly (preferably one sentence per item): geographical position of the investigated area, topography, climate, hydrology, geology, hydrogeology, and other relevant information that might assist in setting up the objectives.

    Pollution situation description with respect to hydrogeological setting and pollution targets; very briefly on: pollution (sources, kind, spreading, potential targets) and other relevant information that might assist in setting up the objectives.

    Results of the previous modelling studies should be referenced. All information sources should be referenced.

  1.2 Management objectives

    Describe MDO and express it eventually in terms of decision variables (state variables or hydrological stresses: GWL, concentration, pumping rates). The role of constraints should be elaborated as well.

  1.3 Modelling objectives

    By answering the question: 'What kind of information is expected from the results of the modelling study?' state MO as precisely as possible.

    Justify the assertion that the modelling is the best way to provide the required information.

    Review all assumptions incorporated within the MO with respect to reality and their potential impacts on MDO.

    State the importance of decisions which will be influenced by the model results and their sensitivity to the range of possible or likely outcomes of the modelling.

    Any change of MO in a course of modelling have to be reported.

  1.4 Model complexity

    Initially MCO outcome of analysis performed during the DP, updated during modelling. The record should be presented, showing when and why MCO has been changed. Information on type of the model, time regime, dimensions, modelled processes, shape complexity, etc. should be briefly presented here. An MCO overview can be given in tabular form with adjacent elaboration (if necessary).

1.5 Miscellaneous

Composition of the modelling team, quality assurance and the peer review process should be described. A financial budget appropriate to the modelling objectives has to be prepared and documented during the DP step; it should not, however, be included in an open-file report.

2. Conceptual Model

2.1 GWS framework (Model matrix and boundaries)

Brief overview of geological units (genesis, lithology, stratigraphy) referring to the regional geology where necessary. Overview of the hydrogeological units, explaining translation from geological information where necessary (especially when distinctive lithostratigraphic units are combined in an aquifer/aquitard or disregarded). Basic attachments:

- *Table* showing chronostratigraphy, geological and hydrogeological units.
- *Maps*: geological (lithostratigraphic) and hydrogeological maps showing 'thickness' of the units and 'depth to' bottom unit (GWS basis).

Selection of GWS boundaries has to be explained (elaborated), using information on geology, hydrogeology, GWL, surface water, potential targets of pollution, management practice, etc.

2.2 Hydrological stresses

2.2.1 Boundary conditions

Brief description of conditions at the system boundaries (upper, lower and lateral). The basic attachment is a *hydrological map* showing:

- surface water courses (indicating those taken (if taken) as GWS boundaries);
- spatial distribution of precipitation and evapotranspiration (if needed),
- GWL hydroisohypses taken as lateral boundaries (if applied).

(GWL hydroisohypses inside the GWL boundaries and pumping/recharge wells should also be included here if the map does not contain too much information - otherwise an additional map should be included in the report section 2.2.2.)

Charts, *tables* and *diagrams* showing temporal distribution of boundary variables are either recommended (as confirmation of the decision to select a time-independent state at the boundaries) or compulsory (if a time-dependent state at the boundaries is selected).

2.2.2   Sinks and sources

Brief description of pumping and/or recharge fields accompanied by a *hydrological map* where locations of wells (and galleries and ponds) are shown (if the hydrological map contain too much data, these locations should be given on separate (additional) *hydrological (II) map* or *'sources and sinks'* map, together with GWL hydroisohypses located inside the GWS boundaries. The same should be repeated for each aquifer where the pumping/recharge is taking place.

A *table* and/or *diagram* showing the temporal distribution of pumping/recharge is either recommended or compulsory (depending on the decision as to whether to introduce constant or time-variable pumping/recharge into the model).

2.2.3   GWS Budget overview

A *budget table* showing all the balance components and the (calculated) sums of input and output components should be presented.

The budget should be calculated (and presented in a tabular form) for several time steps (e.g. to show seasonal variations) if transient modelling is selected.

2.3   Flow and transport in the GWS

2.3.1   Flow in the GWS (flow pattern and flow regime) - GWL (state variable)

The GWS framework (report section *2.1*) provides the basis for the flow pattern. A description of the flow pattern in the current section should be accompanied by a *GWL map* (for each aquifer not covered by 'hydrology' and 'sinks and sources' maps). The groundwater level (head) hydroisohypses exhibit a certain pattern that is used in determination of the general direction of the groundwater flow through the system. The map could also indicate GWS recharge and discharge zones, as well as the relation between aquifers (seepage-infiltration zones). Groundwater (chemical) characteristics can also be used to indicate flow pattern and regime. A considerable part of information for estimation of flow regime should be obtained from report section *2.2* (hydrological stresses).

2.3.2   Transport in the GWS (pattern and duration) - concentration (state variable)

A contaminant *identity table* should be prepared, containing basic information on contaminant(s). That information (concentration, mobility, solubility) should be used to describe (possible) transport of contaminant(s) within the GWS. Information on GWS framework, flow pattern and regime also yields useful information (if a pollutant is transported along the groundwater flow). Impact of management practice on pollution spreading has to be elaborated and documented as well.

The maps showing *pollution source* and *pollution plume* should be prepared (it can be only one map, if the scale of the GWS allows it, or if no data on the pollution plume are available).

2.4 Model Parameters

2.4.1 Flow parameters

Report on basic parameters: permeability, resistivity, porosity and specific yield/storage parameters (if transient modelling). The description should be accompanied by *the maps* (for the spatially distributed parameters) and a unique *table* (for the parameters approximated as spatially uniform)

2.4.2 Transport parameters

Report on basic transport parameters: Solubility or K_{ow} (needed for K_{oc} estimation), dispersivity, diffusivity, organic carbon content, dry bulk density and the first order degradation constant (half-life). Finally, value of estimated retardation coefficient should be reported. The transport parameters can be presented in a unique *table*.

2.5 Miscellaneous GWS characteristics

Groundwater (chemical) characteristics, management practice, other relevant information.

3. Modelcode Selection

Brief description of the selected modelcode, including main characteristics (encoded solution technique, type (flow or flow and transport), regime, dimensions, etc), manufacturer, availability of peer reviews, verification studies, relevant field tests and evidence of modelcode acceptability in the modeller community.

Review of the main factors (related to the problem at hand) that have influenced the selection.

Possible impact of the modelcode assumptions on modelling objectives and model complexity; any discrepancy between required MO (and MCO) and the capabilities of the selected modelcode should be identified and justified. If MO are modified due to limitations of the modelcode, those modifications should be documented.

If the modelcode is modified, the following tests should be performed and the testing methodology and results should be justified:

- reliability testing (with regard to mathematics, mass transport and transfer, hydrogeological system representation, boundary and initial conditions),
- usability evaluation (a peer review and verification study),
- performance testing (comparison of the model results with predetermined benchmarks).

4. Model Design (Construction)

This section reports on model construction, i.e. describes model setup and the model item (estimation and) allocation procedure. During the MD, the MC outcome is 'translated' into the numerical model input, according to the specifics of the modelcode. Accordingly, the report on each item should include description of numerical model specifics (i.e. a way of item handling in the model) and actual item values assigned to the model.

4.1 Model geometry

The model grid should be shown as an overlay on a topographic map at an appropriate scale. Model grid extension (boundaries), spacing and orientation should be discussed and justified, based on model objectives and the conceptual model. If the model is 3-D, describe how the layering is constructed (number, type and thicknesses of layers) in the model, and justify the layering based on the conceptual model.

4.2 Model variables

4.2.1 **Boundary conditions** describe how the external boundaries are constructed in the model and present actual values allocated to the grid cells.

4.2.2 **Sinks and Sources** present locations (in map form) of sinks and sources, their respective stress rates, and how they are incorporated in the model.

4.2.3 Initial conditions

Groundwater level/head values have to be assigned to each grid cell. The method used for extrapolation of observed/measured data should be described. Initial contaminant concentration for transport modelling should also be reported.

4.3 Model Parameters (including data averaging and/or estimation procedure)

Flow, transport and time parameter values should be presented. If parameters values vary spatially in the model, present this distribution as a map (refer to the conceptual model). The data averaging and/or estimation procedure used in model construction should be described.

5. Calibration

The report on the calibration process begins with description of calibration targets and corresponding acceptable residuals. Sources and magnitudes of errors associated with each calibration target value should be reported. Calibration parameters should be listed, and their selected justified. The following should be prepared:

– a map showing location of calibration targets relative to nodes in the grid, and
– a table showing initial parameter estimates and their coefficients of variation.

5.1 Residual analysis and model verification

The procedure for history matching should be described, including a description of approach used and 'calibration sensitivity analysis' performed. The qualitative and/or quantitative tests used in history matching should be described. The rationale for selecting or omitting the comparison test should also be stated. Eventually, the results of comparison, i.e. residual analysis, should be presented in a systematic way. Spatial distribution of residuals can be presented in several ways, such as:

- a *map* of superimposed contours of head,
- a *map* showing contours of head residuals,
- a *map* showing location and value of calibration targets and simulated values,

- a *plot* of calibration values vs. simulated values, showing deviation from a straight-line correspondence,
- a *box plot* of residual heads for each calibration run,
- *plot* of Mean Error (ME), Mean Absolute Error (MAE) and Root Mean Squared (RMS) error vs. the calibration run number to show the approach to calibration,
- *plot* of ME, MAE and RMS vs. parameter values to show the sensitivity of the calibration to changes in a parameter value.

The results of model verification performed with independent data sets should be reported; the match between field and simulated heads and fluxes should be presented in a same way as that used to present calibration results.

5.2   Sensitivity analysis

If a sensitivity analysis is performed, the report should state which model inputs were varied and which computed outputs were examined. The report should justify the selection of model inputs and computed outputs in terms of the modelling objectives.

For each model input that was varied, the report should present a graph showing the changes in residuals (or residual statistics) and the computed outputs with respect to changes in the model input. The report should either state that none of the analyses had a Type IV result, or else identify which analyses had Type IV results (see ASTM, 1994c).

If a sensitivity analysis is not performed, the report should state why a sensitivity analysis was not needed. If the ASTM approach is adopted, the sensitivity analysis should be performed after predictive simulation (see section 6.3.6).

6.   Predictive simulations

The differences (in input parameters and simulation results) between predictive runs should be described and presented in tabular form. Sensitivity analysis should be performed for at least one of the predictive simulations (predictive sensitivity analysis). The results of predictive modelling can be presented by using *diagrams* and *contour maps*.

7.   Summary and Conclusions

Include an Executive Summary (in terms of the decision objectives). The success of the model application in simulating the site scenarios should be assessed. The assessment should include an analysis of (OSWER, 1994):

- whether the modelling simulations were realistic,
- whether the simulations accurately reflected the scenarios
- whether the hydrogeological system was accurately simulated
- which aspects of the conceptual model were successfully modelled.

Possibilities to conduct postaudit and merits of postaudit should be discussed.

Remark 1: Postaudit is usually reported separately.

Remark 2: The Quality Assurance (QA) process established at the beginning of the project (see the report item *1.5*) should be carried out (by independent reviewers) and reported in a form of a 'peer review'. A peer review is prepared as a separate report. It reviews basically: modelling objectives development, conceptual model development, modelcode selection, model setup and calibration, simulation of scenarios and postsimulation analysis (postsimulation analysis is here considered as a part of conclusion/completion process).

6.4   Knowledge systematisation and formalisation

Ordering of modelling tasks was the first taxonomical task, i.e. the first step in knowledge systematisation. Subsequent acquisition of knowledge meant, at the same time, further elaboration (systematisation) of the modelling tasks. The part of systematisation that follows knowledge acquisition is primarily related to the transfer of knowledge to the user. For electronic encapsulation some form of knowledge representation (as offered by Artificial Intelligence) needs to be selected and implemented.

Various forms of knowledge representation have been applied (or are intended to be applied) in GMM development. Analysis of modelling tasks and knowledge acquisition revealed rather poor taxonomy and a lack of formalised expertise. Consequently, only a small portion of knowledge could have been expressed in the form of straightforward rules. Further systematisation of acquired knowledge showed that the module should be organised as an 'electronic' modelling protocol. General knowledge related to the modelling protocol was formalised in hypertext *topics*. Specific knowledge on model complexity was expressed in *rules* and organised in an independent rule-based module (MCM). It has also been found that knowledge on parameter estimation should be organised (and encapsulated) in parallel with the related numerical information in independent modules (RE1 and DE1). Finally, it is the intention to encapsulate modelling case-studies electronically and to use a *Case-Based Reasoner* to control their storage and retrieval.

Topics, hypertext-based knowledge representation forms, were already introduced in the Site Characterisation Module (Chapter 4). GMM topics were structured basically in the same manner as SCM topics. They contain general knowledge that can be used as a basis for further structuring and formalisation of encapsulated knowledge. As already mentioned (in Section 3.5), hints, advice and recommendations can be 'extracted' (in the next stage of GMM development) from the topics and 'transformed' into rules. Some of the 'potential rules' are already shown in topics as underlined text, in order to attract user attention (and alleviate intended extraction). A few examples are given below:

– A two order of magnitude contrast in hydraulic conductivity may be sufficient to justify placement of an impermeable boundary;

– The effects of the boundary conditions may be tested by changing specified head conditions to specified flux and vice versa;

- The boundaries of the model should be located along natural limits if possible, or well away from the region affected by pumpage or recharge;

- Steady-state problems require at least one boundary node with a specified head in order to give the model a reference elevation from which to calculate heads;

- Whenever possible, specified head conditions are selected over specified flow, because... In some situations, however, it may be advisable to use specified flow conditions. For example, flows into a system may be constant whereas heads along the boundary may be expected to change during the simulation.

Some additional features (with respect to the SCM) were introduced in the GMM to facilitate flow in the module. They are the result of systematisation carried out in accordance with the specifics of groundwater modelling knowledge. (Examples of GMM topics and related features will be given in the next section.) Special attention was, however, paid to systematisation of MC items, as explained in Section 6.3.2. SC yields input for determination of MC items; for example, one of the MC inputs is groundwater level/head (Figure 6.3). The modeller usually approaches conceptualisation from the side of MC items. Table 6.2 (a table segment is shown in Figure 6.3) contains a list of MC items that request (and use) each particular SC outcome (e.g. groundwater level/head); in the current example, definition of 'Boundary Conditions' requests (among other SC outcomes) information on groundwater level/head. By activating Model Conceptualisation Chart (Figure 6.2 and Figure 6.3), the user obtains (in a pop-up window) a list of requested SC outcomes for each particular MC item (thus, the other way around), and a direct link with the outcomes (presented in Table 6.2). It is not claimed here that either tables or lists (charts) are complete. However, the approach is certainly worthy of further elaboration.

The *production rules* are the most widely-used forms of knowledge representation, often organised in so-called decision tables, or in more complex structures (for the overview, see Section 2.4). The rules on groundwater modelling acquired to date are on general model complexity; model complexity framework is expressed through decisions such as those on type of model, time regime, model dimensions, boundary conditions, pollution situation, etc. These decisions are taken by the modeller each time at the beginning of the modelling process (i.e. Define Purpose), although some of them can be revised during modelling. The acquired rules have been encapsulated in the rule-based Model Complexity Module (a description of MCM is given in the next section).

Most of encapsulated rules were extracted from KGM (Section 2.2 and Section 2.3.2), and others are suggested by modelling experts.Only those entries that showed relevance to the chosen scope of the Modelling Protocol were extracted from KGM. Firstly, the KGM was run for an example of a point-source pollution problem that takes place in (non-consolidated sedimentary) porous saturated medium. Subsequently, all the KGM entries (48 of them for the specified problem) were analysed, as exemplified in Table 6.4. The analysis yielded a set of entries (decision tables)

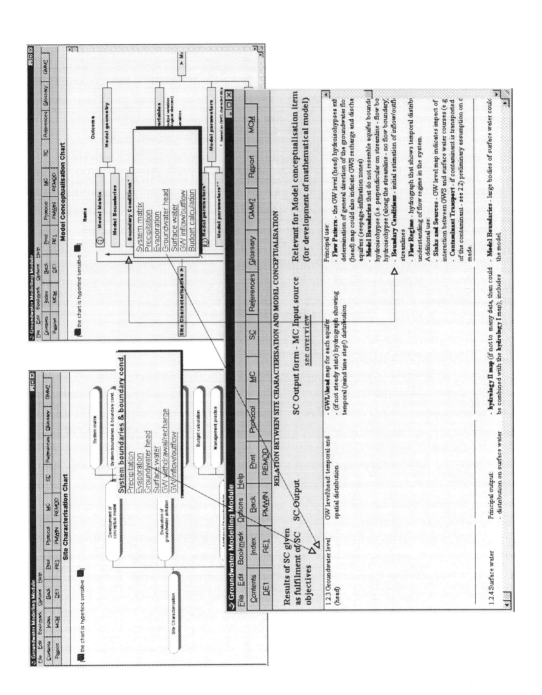

Figure 6.3 Connecting MC items with SC results

that have been included in the MCM. An example of a selected entry is given in Table 6.5.

Table 6.4 Analysis of the KGM entries (example)

Entry	Options	Purpose/Comment
Broad problem analysis		
– What is the source of contamination?	air pollution, agriculture, polluted river, waste disposal site, spills, injection wells	forward suggestion for 'cont. types' forward suggestion for 'duration'
– What is (are) the type(s) of contaminant?	P nutrients, N nutrients, cation salts, anion salts, heavy metals, cyanides, radionuclides, oil products, pesticides, VOC, BOD	forward suggestion for 'phase' forward suggestions for 'reactive processes' ('presence' and 'type')
– What is the zone of the problem?	*saturated, unsaturated*	*abandoned - not relevant*
– What is the landform? Replaced with: What is the landscape?	mountains, hills, polder, wetland, plain, island, cliff, river bank	final suggestion for 'dimensions' (mountains and hills: fully 3 D) if fully 3D - anisotropy suggested
– What is the rock type?	*non-consolidated sediment, consolidated, volcanic, plutonic, metamorphic*	*abandoned - not relevant (only non-consolidated sediments are considered)*
– What is the type of sediment?	alluvial fan, eolian, coastal plane, glacial, tectonic valley	forward suggestion for 'texture' final suggestion for 'heterogeneity' final suggestion for 'head dependency'
– What is the climate?	*cooler-humid*	*abandoned - too broad (e.g. suggestion for occurrence of surface water)*
– Which type of results are required?	flow, flow + transport	forward suggestion for 'processes' involved in suggestion for 'time regime'
– Which accuracy is required?	rough, precise	involved in several suggestions
– What is the spatial scale?	*local, subregional, regional*	*abandoned - not relevant*
– What is the temporal scale?	short term , long term	final suggestion for 'time regime' (very weak!)
Detailed problem analysis		
– What is the faze of contaminant?	solid, dissolved, gas, NAPL	final suggestion for 'effects on flow system'
– What is duration of contamination?	instantaneous, permanent, time dependent	final suggestion for 'transport time dependency of the boundary'
(continued up to entry 48)		

Postulates of *case-based reasoning* were presented in Section 2.4.2. Prerequisites for application of CBR are quality (accuracy) and consistency of cases studies to be stored in the knowledge base. The GMM has been developed to enhance accuracy of modelling; The GMM also contains the recommended content of modelling report (Section 6.3.6), that should, together with the REPORTER formatting template, secure consistency in reporting. The potential place and role

Table 6.5 Selected KGM entry, an example

source type	waste disposal sites	spills	injection wells	polluted surface water
suggested contaminants	N nutrients && P nutrients && anion salts && cation salts && heavy metals && cyanides && radionuclides && oil products && VOC && BOD	N nutrients && P nutrients && anion salts && cation salts && heavy metals && cyanides && pesticides && oil products && VOC && BOD	N nutrients && P nutrients && anion salts && cation salts && heavy metals && cyanides && pesticides && radionuclides && oil products && VOC && BOD	N nutrients && P nutrients && anion salts && cation salts && pesticides

of CBR in DSS for Groundwater Pollution Assessment will be discussed in the next section.

6.5 Software development and integration

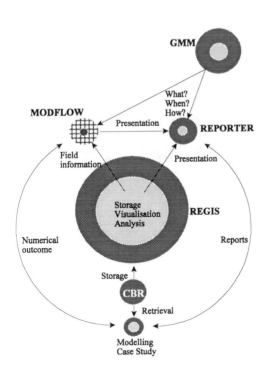

Figure 6.4 DSS components to be used in modelling

GMM software has been developed to enhance the modelling procedure and, together with REPORTER, to secure consistency of modelling reports. Modelling reports (i.e. modelling case-studies) will be used by CBR in order to extract knowledge useful for a new groundwater modelling problem. The place of the GMM, REPORTER and CBR in the DSS for Groundwater Pollution Assessment is shown in Figure 6.4.

The role of the GMM is to assist a modeller in conducting the modelling procedure (what? when? and how?). Field information stored in REGIS is used to prepare input for MODFLOW, the modular 3D finite difference groundwater model (McDonald and Harbaugh, 1988). The MODFLOW input is prepared partly in REGIS (a model grid and spatially distributed, grid-related input data) and partly in PMWIN or in the REMOD module (see Section 3.4). Modelling reports are written (with assistance of GMM) in a text processor

(REPORTER) that provides a predefined reporting format. The modelling report and model numerical outcome make up a modelling case-study (Figure 6.4). CBR controls storage and retrieval of case-studies.

GMM software, REPORTER and the modelling environment (REGIS, PMWIN and REMOD) will be described in more detail below.

6.5.1 GMM software

The GMM consists of several software applications developed in various programming environments. The core of the GMM is an electronic Modelling Protocol guidance composed of sets of hypertext-based topics. The knowledge on general model complexity is encapsulated in a rule-based Model Complexity Module (MCM). The GMM also contains RE1 and DE1 modules that assist in estimation of the retardation factor and dispersivity, respectively. The

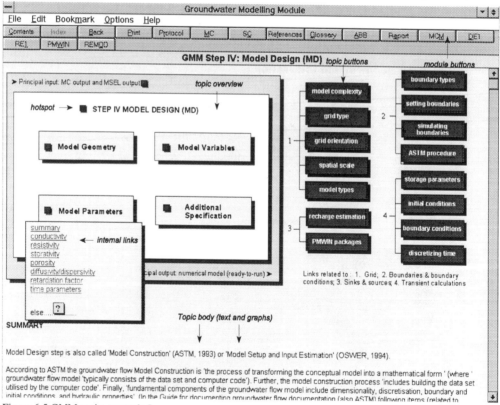

Figure 6.5 GMM topic, example I

Modelling Protocol acts as a platform that integrates the software applications in a unique DSS module.

The main GMM window contains the *Modelling Protocol* chart (Figure 6.1) that links modelling steps with corresponding sets of topics. Each topic (e.g. Figure 6.5) contains external and internal links. *External* links are established with other related topics (i.e. previous or subsequent modelling steps, references, glossary, or topics containing more elaborated knowledge) and other

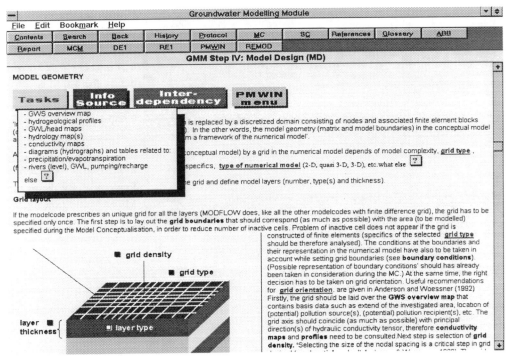

Figure 6.6 GMM topic, example II

software. The main external links are gathered as 'topic buttons' and 'module buttons' at the top of the topic and the GMM window, respectively (Figure 6.5). *Internal* links connect the overview of the main tasks of a particular modelling step, given at the top of the topic, with corresponding sections in the topic body. These sections have often undergone further systematisation. For example, by choosing 'Model Geometry' (and than 'summary') in the topic overview, detailed information on Model Geometry will appear (Figure 6.6). The (sub)tasks related to the definition of Model Geometry are summarised at the top (button 'Tasks' opens a pop-up window) and described in detail further in the text. Main information sources (MC results) are also summarised (this pop-up window is shown in Figure 6.6) and marked darker (bold) in the text. Additionally, an attempt has been made to summarise relations between a current tasks and other modelling tasks ('inter-dependency' button); for example, time discretisation should be taken into account while designing the grid (i.e. defining spatial discretisation) if transient modelling is planned. Connections with the corresponding part of the PMWIN menu can be established via the 'PMWIN menu' button, while the topics giving more information on the Model Geometry can be reached via links embedded in the text (highlighted, underlined text); these are the very same links gathered as 'topic buttons' at the top of the topic (Figure 6.5). In total, the GMM prototype consists of about 250 topics, mutually connected by about 350 links.

Model Complexity Module (MCM)

The MCM is a rule-based module developed in a Knowledge Base Editor (KBE).[15] Modelling

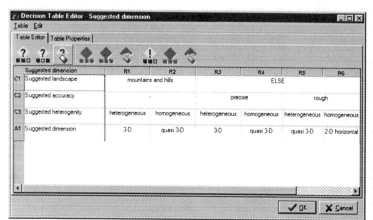

Objectives have to be defined and the Summary of Site Characterisation outcome has to be known before the MCM is used. MCM outcome is the model complexity framework, expressed through decisions taken during the first modelling step ('Define Purpose'). By using the MCM the modeller approaches the issue of model complexity in a

Figure 6.7. An example of a MCM decision table

systematic and consistent manner; moreover the MCM takes care that none of the decisions (regardless how simple they might be) are left out (forgotten).

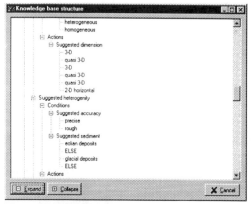

The main qualities of MCM software are transparency and adaptability. All the decision tables (that contain encoded rules) are fully transparent (Figure 6.7). The modeller can modify rules, as well as add new rules to the MCM. A decision tree (Figure 6.8) contributes to transparency of the inference procedure and, accordingly, alleviates MCM modification and/or enlargement.

The MCM user interface is very simple; the user is requested to answer a set of questions, by choosing one of the offered options. This process is supported by a kind of on-line help

Figure 6.8 MCM decision tree

where the options are textually or graphically explained (the on-line help can be modified as

---

[15] Knowledge Base Editor is a product of TNO Building and Construction Research, The Netherlands (no publications available).

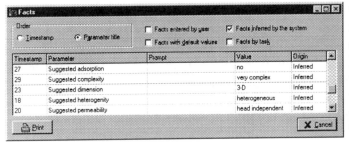

Figure 6.9 An example of a MCM fact table

well). Finally, the user has a full insight into the origin of decisions made in t he module MCM It can be clearly seen which decision are inferred by the MCM (Figure 6.9), where the inference is done ('facts by task' - Figure 6.9 and Figure 6.8) and why (Figure 6.7).

The MCM is seen as an example of encapsulation of rule-based knowledge in groundwater quality modelling. Availability of transparent and adaptable knowledge editors makes encapsulation easier, more relevant and attractive.

Dispersivity parameter module (DE1)

Model Complexity analysis should point out whether dispersivity needs to be considered in the model. Extensive information on dispersivity (and related diffusivity) is stored in the Site Characterisation Module (Chapter 4), next to the information on other processes that control mass transport and transfer in the groundwater system.

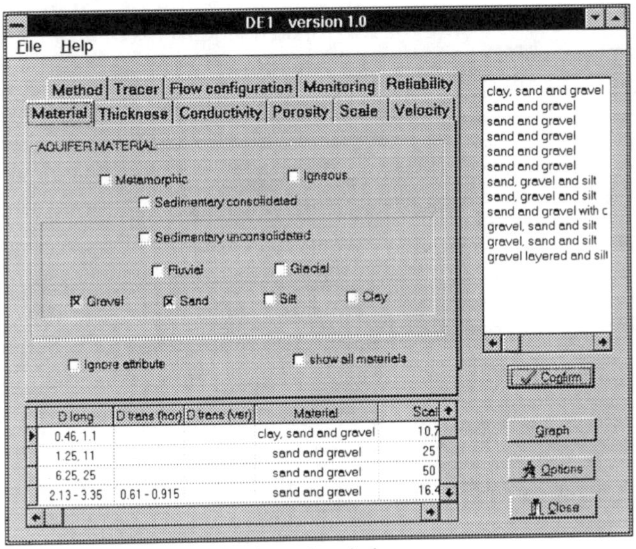

Once a decision is taken to consider dispersivity in the model, the DE1 module can be used to assist in selection of dispersivity values. The DE1 database contains information on dispersivity collected from 59 different sites (Gelhar et al, 1992). Besides sole dispersivity values, the database includes a number of attributes (basically on site and method used) that might be useful in selection (Figure 6.10).[16] In most of the cases determination of attributes is not straightforward, mainly owing to the variety of attribute descriptions. Therefore, a number

Figure 6.10 DE1 moodule; the main window

---

[16] In 'Envirobrowser' software (GEOREF, 1997) selection of dispersivity is based only on type of aquifer material; the DE1 was developed because it was found important to include additional attributes (besides aquifer material) in the database.

of options is created to avoid rigidity in attribute determination; the most common attribute descriptions are shown in the user interface to be selected individually or combined (e.g. gravel and sand, fluvial and glacial, etc). The user can also list all attribute descriptions available in the database and determine attributes directly from the list. Eventually, the user can disregard attributes that are not found important for selection of dispersivity values.

The DE1 also contains a well-known graph that shows longitudinal dispersivity versus test scale (for various methods and aquifer materials). An additional option should be developed in the DE1 prototype to allow database enlargement by the user (this option is available in the MCM).

Retardation factor module (RE1)

The RE1 module assists in estimation of the retardation factor. If retardation due to sorption is found important enough (Model Complexity) to be included in the numerical model, the retardation factor has to be estimated. As already mentioned, PMWIN also calculates retardation factor, so RE1 can be used to define K_d, the linear sorption coefficient required for this calculation. K_d is determined from K_{oc}, the organic carbon/water partition coefficient and f_{oc}, the organic carbon content of the stratum material. Solubility values or K_{ow} values (octanol-water partition coefficients) can be used in the definition of K_{oc}.

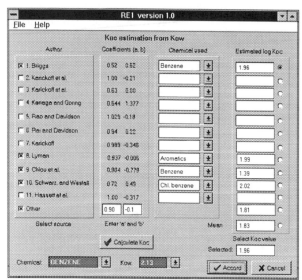

One of the RE1 screens is shown in Figure 6.11. A number of researchers have found relationships between the K_{ow} and K_{oc} values for various organic compounds. Accordingly, equations have been derived, and the modeller is placed in the situation of deciding which equation to use. The logical choice would be the equation that is derived on the basis of chemicals similar to the one under study. RE1 provides the user with the option to make this selection, to compare values obtained by various equations and to calculate K_{oc} by introducing a new linear relationship. The module could be extended to

Figure 6.11. An example of a RE1 window

support description of equilibrium by Freundlich and Langmuir isotherms (besides the linear one).

RE1 and DE1 are stand-alone applications developed as examples of modules that can assist in selection or estimation of various parameters required in assessment of groundwater pollution.

6.5.2 Reporter

The case specific nature of groundwater pollution problems obstructs (hinders) modellers in applying a uniform approach to reporting on modelling results. At the same time, it gives them the freedom in reporting, which some modellers use as an invalid excuse for not trying to be consistent. An additional problem is a 'personal writing style'. Technical writing is a skill that can be learned, but expressing is still a matter of personal eloquence. Finally, as reporting means conveying information (to potential readers), modellers should be constantly aware (while reporting) that the report needs to be intelligible to the others, and not just to the author of the report.

The reporting procedure needs to be established already at the beginning of the modelling, and

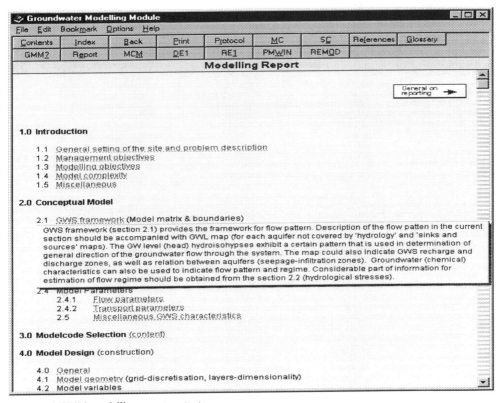

Figure 6.12 GMM: modelling report content

carried out during every modelling step (see Section 6.2). The report is finalised during Report Completion (Figure 6.1), after 'overall effectiveness' is evaluated. The GMM contains the general content of a modelling report (Section 6.3.6). Each reporting item in the GMM is accompanied by a pop-up window (Figure 6.12). Pop-ups contain information on items that should provide answers to the following questions:

- what needs to be explained (and written down)?
- what are (possible) links among the items? and
- what kind of graphical/tabular representation is the most appropriate?

In order to secure consistency in reporting, a template has been made in the Word text processor

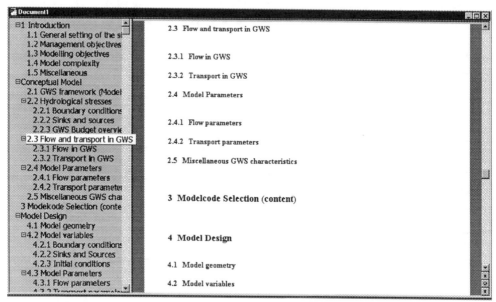

Figure 6.13 REPORTER (Word template)

that contains the same items given in the GMM (Figure 6.13). The template is based on NITG-TNO (Netherlands Institute of Applied Geoscience) 'house-style' reporting template. Development of a new text processor, i.e. REPORTER meant only for modelling, has not been found necessary at this stage. Text processors like Word Perfect and Word are widely used, and development of a comprehensive text processor would be a complex and time consuming task. However, decision on development of new REPORTER needs to be reconsidered in the next phase of this project. Modelling reports will serve as inputs for a case-based reasoner and it is not yet decided what kind of requirements might be imposed on reporting format. For instance, an option might be needed for preliminary selection of CBR indexes already during the reporting.

An additional issue is graphical/tabular representation of modelling results. Modelling results are not only textually described, but also presented graphically (maps, diagrams) and in tables. Figures and tables prepared in PMWIN or REGIS should, together with the textual part of the report, be stored in REGIS (see Figure 6.4). Instead of saving graphs that are memory consuming, queries (sets of predefined commands) should, whenever possible, be created and attached to the text. The queries can afterwards be used to (re)produce graphics whenever they might be needed. REPORTER should be able to recognise and store this kind of information.

6.5.3 Modelling environment

The DSS components to be used in modelling were shown in Figure 6.4. Data required for the modelling are stored in REGIS, whereas the actual data processing is carried out by MODFLOW. Accordingly, integration of these two DSS components was one of the priority tasks in development of the DSS for Groundwater Pollution Assessment. The coupling between REGIS and MODFLOW is fully operational, being established via PMWIN and via REMOD (Figure 6.14).[17] A MODFLOW grid and data set prepared in REGIS are retrieved in PMWIN or REMOD in order to prepare MODFLOW input files. Visualisation of modelling results and their storage are provided by REGIS (for details see Kukurić, 1998).

Preparation of a MODFLOW grid and the grid-like data set in REGIS consist of several simple steps (Figure 6.14):

- selection of a model area

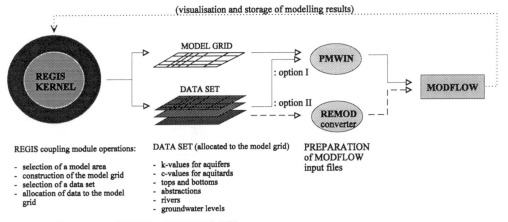

Figure 6.14 Integration of REGIS with MODFLOW

By using standard REGIS options the user can position the REGIS screen on the area of interest (a little larger area should be chosen). Then those features should be shown on the screen (aquifer parameters, wells, rivers, etc) that might influence grid construction (Figure 6.15).

- construction of the model grid

---

[17]PMWIN is a pre- and post-processing shell around several modelcodes, including MODFLOW (see also Section 3.4). REMOD is an independent module that, like PMWIN, converts REGIS output files into MODFLOW input format. Unlike PMWIN, REMOD does not have a Graphical User Interface (GUI) for preparation/modification of spatially variable data files and for visualisation of MODFLOW results.

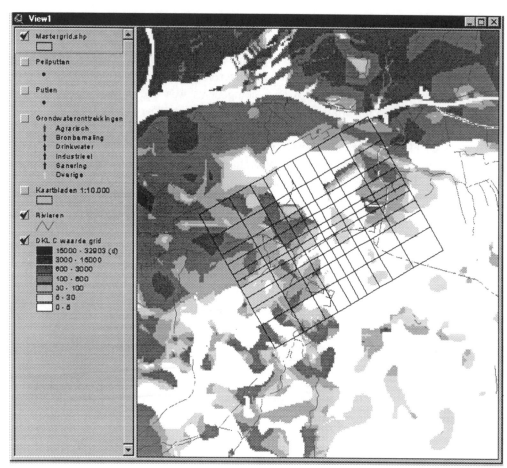

Figure 6.15 Selection of model area in REGIS

By selecting 'MODFLOW' on the main REGIS menu, a drop-down menu will appear, having three items: Project, Layers and Generate. If 'Project' is chosen, a new window (Figure 6.16) will appear next to the main REGIS screen. The project name has to be specified firste; the model grid and files to be generated will be saved in a sub-directory having the project name. This option allows convenient re(use) of the generated files. Model grid area can then be selected, either by using the mouse or by typing the zero coordinates, columns width and rows height (Figure 6.16).

- selection of a data set

By selecting 'Layers' from the drop-down menu, a window will appear with the list of hydrogeological layers present within the grid area. Once the layers are selected a drop-down item 'Generate' (Main REGIS menu) should be chosen.

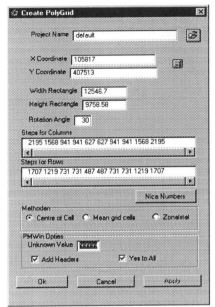

Figure 6.16 Creation of a model grid

- allocation of data to the model grid

REGIS will automatically generate a grid file and data files with aquifer/aquitard parameters for each selected layer. Generated data files contain information on: aquifer conductivity, aquitard resistivity, layer top and bottom (thickness), river parameters, withdrawals and groundwater levels. The grid file has the same format as a file that is created when a grid is prepared in PMWIN. One of the PMWIN initialisation files is also modified in REGIS according to specifics of the newly-constructed grid. The grid file and the altered initialisation file are copied to the project sub-directory together with data files. PMWIN uses information from these two files to construct the grid. Data files need to be retrieved one by one for each parameter. After that the grid and parameter files can be modified, and MODFLOW can be run.

Data files prepared in REGIS can also be retrieved in REMOD in order to prepare MODFLOW input files. REMOD is developed for modellers who do not use commercial (PMWIN-like)

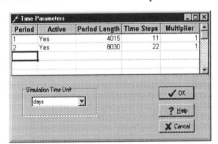

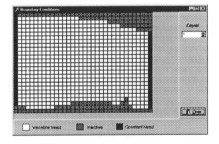

Figure 6.17 Examples of REMOD interface

shells. REMOD allows preparation of the MODFLOW packages (BAS,BCF, RIV1, etc) one-by-one, meaning that not all input packages need to be (re)generated if only one package requires alteration. On the other hand, if the packages are prepared together (during one REMOD run), than REMOD automatically passes common information from one package to the other. Examples of the REMOD interface are shown in picture 6.17.

7. CONCLUDING REMARKS

Electronic Decision Support Systems are (or should be) integrated information systems. They consist of software tools developed for the storage, processing and presentation (and interpretation) of information. DSSs should be conceptualised in such a way as to provide the optimal flow of information through the system while performing user-assigned tasks. Both quantitative (numerical information, data) and qualitative information (knowledge) are needed to support decision-making. To date, most of attention has been paid to numerical information, and too little to knowledge.

The research described in this thesis resulted in the development of a DSS for groundwater pollution assessment. The DSS is a prototype of a task-oriented integrated information system. It is meant to assist the user in carrying out the following tasks: site characterisation, vulnerability assessment and groundwater pollution modelling. For each of these tasks Knowledge-Based Modules (KBMs) were developed and integrated into the DSS. The modules contain a substantial portion of knowledge on groundwater pollution problems in terms of parameters, methods and procedural steps required for accomplishing the tasks. They also contain embedded links with the software that stores, processes and presents numerical information within the DSS. In short, the KBMs can be used as a knowledge repository, as a guide, and as an executive DSS interface.

The modules were developed from scratch, applying a knowledge encapsulation procedure. Electronic encapsulation of knowledge is a complex process that includes: definition of the problem, knowledge acquisition, systematisation, formalisation, design and development of a knowledge component and its integration into a DSS. Most of the effort was put into the first three steps of the procedure. The basic taxonomy of the site characterisation task needed to be worked out prior to any knowledge acquisition. Subsequent acquisition of necessity very extensive because some parts of the encapsulated knowledge (on parameters, parameters and processes) are also used during the vulnerability assessment and groundwater modelling. Due to the limited time available for the research, only a few examples of site characterisation processing and presentation were worked out. Further development of this module should concentrate on these parts, including definition of procedures (for processing and presentation) and their coupling with corresponding DSS software.

A newly-developed methodology encapsulated in the KBM for vulnerability assessment incorporated the best features of several methodologies previously developed in this field. In that sense implementation of the encapsulation procedure was much less demanding than in the case of site characterisation. Thanks to that, the time was made available for an intensive testing of the encapsulated methodology. The module (software) was also tested by using a (hypothetical) case study. Further testing should include several real-world case studies in order to evaluate module capability for comparison of groundwater vulnerabilities from different sites.

The protocol encapsulated in the KBM for groundwater pollution modelling is actually an electronic modelling guidance that provides links with a groundwater model and other DSS software required to accomplish a modelling task. Although the protocol covers all the modelling steps, most of the time and effort was dedicated to the model conceptualisation. This very important modelling step have not been sufficiently addressed in either practice or in research circles. Next to the modelling protocol, the KBM for groundwater pollution modelling contains rule-based expert knowledge on modelling complexity and two examples of knowledge encapsulation for the purpose of parameter definition. More expert-heuristic knowledge is expected to be found in the deeper domains of some modelling steps, especially in the model design and model calibration steps. These are rather narrow, well-defined knowledge domains, where a 'classical' knowledge acquisition procedure (as developed by Artificial Intelligence) could be applied. That was, unfortunately, not possible during this research where primarily common, general knowledge was dealt with.

The case-specific nature and interdisciplinary character of groundwater problems are the main specific obstacles to more efficient encapsulation of knowledge in the field of groundwater management. Knowledge to be encapsulated is, due to these and other (general) obstacles, often incomplete, inconsistent, poorly structured, and is lacking in formalised expertise. Serious work on the taxonomy of groundwater management tasks and related knowledge is needed; that endeavour asks for involvement and close cooperation of experts from various fields. A strong emphasis should therefore be placed on teamwork, in which pragmatism and consensus need to be adopted as primary guidelines. In that context, some recent initiatives to publish expert's 'rules-of-thumb' (in Dutch magazine 'Stomingen') rand to expose them to open discussion are very welcomed.

Once acquired and systematised (to a certain extent - encapsulation is continuous process), knowledge needs to be 'translated' into an electronic form. Most of the forms of knowledge representation, as offered by Artificial Intelligence, are rule-based. If knowledge can be expressed in rules (and more complex structures - up to objects), these forms should be used. Otherwise, hypertext-based technology might be applied, as was done in this research. New mark-up languages (e.g. XML), recently introduced on the market, provide much better possibilities for encapsulation of large quantities of semi-structured knowledge than 'classical' HTML used in the research (fortunately, conversion is possible). Combined with Java applets (Java is definitely taking over from other programming environments), they create Intranet/Extranet worlds.

The importance of flexible, adaptive, 'learning' software has been repeated several times in the thesis. Thanks to these characteristics, software becomes 'interactive' in the real sense of the term; not only is the encapsulated knowledge transferred to the user, but also the user has the opportunity to transfer her/his knowledge into the computer and improve and augment encapsulated knowledge. This becomes even more important in combination with Internet and Internet-related Information and Communication Technology (ICT) techniques; they make encapsulated knowledge available to an enormously wide circle of potential users/developers.

Users and developers are becoming the same.

Advanced communication and active participation are crucial for 'knowledge processing', i.e. improvement and effective (re)use of already encapsulated knowledge. In the last couple of years the amount of information on groundwater pollution problems available in an electronic form has rapidly increased. New types of software applications are emerging to deal with systematisation of poorly structured electronic information (e.g terminology management software tools). Still, the success of knowledge processing depends substantially on the quality of the originally encapsulated knowledge; a Case-Based Reasoner, for instance, cannot be successful if the encapsulated cases contain inadequate, or inadequately-represented knowledge. CBR and similar approaches have been found attractive (not because of their search engines, but) because they could contribute to proper knowledge encapsulation. A major research project dedicated to CBR has recently started at NITG-TNO.

Knowledge discovery, knowledge engineering, knowledge management, --- we are constantly under a barrage of new 'knowledge' terms. People have dealt with knowledge since the dawn of civilisation. Nowadays they have computers to assist them. How? This was a small contribution to the answer.

REFERENCES

Aamodt, A. and E. Plaza, 1994. Case-Based Reasoning: Foundational Issues, Methodological Variations and System Approaches, *AICOM*, vol. **7**, no. 1, pp. 39-59.

Abbott, M.B., 1991. *Hydroinformatics, Information Technology and the aquatic Environment*, Avebury Technical, Aldershot, UK and Brookfield, USA.

Abbott, M.B., 1992. The theory of the hydrologic model, or: the struggle for the soul of hydrology, in *Topics in Theoretical Hydrology: A tribute to Jim Dooge*, J.P. O'Kane (ed.), Elsevier, pp. 235-254.

Abbott, M.B., 1993. The electronic encapsulation of knowledge in hydraulic, hydrology and water resources, *Advances in Water Resources*, 16, pp. 21-39.

Adams, B. and S.S.D. Foster, 1992. Land-surface zoning for groundwater protection. *J. Institution of Water and Environment Management*, no. 6, pp. 312-320.

Ahmad, K., 1995. Pragmatics of Specialist Terms and Terminology Management, in *Machine translation and the lexicon*, P. Steffens (ed.), Springer, Germany.

Aller, L.A., Bennett, T., Lehr, J.H., Petty, R.J., and G. Hackett, 1987. *DRASTIC: A Standardised System for Evaluating Ground Water Pollution Potential using Hydrologic Settings*, EPA/600-2-87-035, Washington DC, pp. 485.

Anderson, M.P. and W. W. Woessner, 1992. *Applied Groundwater Modelling, Simulation of Flow and Advective Transport*, Academic Press, San Diego, Ca.

ASTM, 1993a. *Standard guide for Application of a Ground-Water Flow Model to a Site-Specific Problem*, ASTM doc. D5447-93.

ASTM, 1993b. *Standard guide for Comparing Ground-Water Flow Model Simulations to Site-Specific Information*, ASTM doc. D5490-93b.

ASTM, 1994a. *Standard guide for Defining Boundary Conditions in Ground-Water Flow Modeling*, ASTM doc. D5609-94.

ASTM, 1994b. *Standard guide for Defining Initial Conditions in Ground-Water Flow Modeling*, ASTM doc. D5610-94.

ASTM, 1994c. *Standard guide for Conducting a Sensitivity Analysis for a Ground-Water Flow Model Application*, ASTM doc. D5611-94.

ASTM, 1995a. *Standard guide for Subsurface Flow and Transport Modeling*, ASTM doc. D5880-95.

ASTM, 1995b. *Standard guide for Documenting a Ground-Water Flow Model Application*, ASTM doc. D5718-95.

ASTM, 1996. *Standard guide for Calibrating a Ground-Water Flow Model Application*, ASTM doc. D5981-96.

Avouris, N.M., 1995. Cooperating knowledge-based systems for environmental decision support *Knowledge-Based Systems*, vol. 8, no. 1.

Bachmat, Y. and M. Collin, 1987. Mapping to assess groundwater vulnerability to pollution, in *Vulnerability of soil and groundwater to pollution*, W.van Duijvenboden and H.G. van Waegeningh, (eds.), TNO Committee on Hydrological Research, The Hague, Proceedings and Information No. 38, pp. 297-307.

Bakoyni, P., 1993. Expert Systems in Hydrology, *Technical Reports in Hydrology and Water Resources*, no. 36, WMO Geneva.

Bear, J. and A. Verruijt, 1987. *Modelling groundwater flow and pollution*, Reidel Publishing Co, Dordrecht.

Bear, J., Beljin, M.S. and R.R. Ross, 1992. *Fundamentals of groundwater modelling*, US EPA, EPA/540/S-92/005.

Bedient, P.B., Rifai, H.S. and C.J. Newell, 1994. *Groundwater Contamination and Remediation*, Prentice- Hall.

Bell, D., 1973. *The Coming of Post-industrial Society: A Venture in Social Forecasting*, Basic Books, New York.

Bellman, R.E., 1978. *An Introduction to Artificial Intelligence: Can computer Think?*, Boyd & Fraser Publishing Company, San Francisco.

Bieshuvel, A. and C.J. Hemker, 1993. Groundwater modelling and GIS: integrating MICRO-FEM and ILWIS, in *Application of GIS in Hydrology and Water Resources*, K. Kovar and H.P. Nachtnebel (eds.), pp. 289-296, IAHS Publication 211.

Brusseau, M.L., 1994. Transport of reactive contaminants in heterogeneous porous media, *Reviews in Geophysics* 32 (3), pp. 285-313.

Carrera, J. and S.P. Neuman, 1986. Estimation of aquifer parameters under transient and steady state conditions, *Water Resources Research* 22(2), pp. 190-242.

Castells, M.., 1996. *The Rise of the Network Society*, Blackwell Publishers, Oxford UK.

Charbeneau, R.J., Bedient, P.B. and R.C. Loehr (eds.), 1992. *Groundwater Remediation*, Water Quality Management Library, vol. 8, Tehnomic publishing.

Chiang, W-H. and W. Kinzelbach, 1996. *Processing Modflow for Windows, software and manual*, distributed by Scientific Software Group, USA.

Connor, J.A., 1994. Hydrogeologic Site Investigations, *GroundWater Contamination* by P.B. Bedient, H.S. Rifai and C.J. Newel, Prentice Hall.

Crowe, A.S. and J.P. Mutch, 1994. An expert Systems approach for assessing the potential for pesticide contamination of ground water, *Ground Water* vol. 32, no. 3.

Datta, B., Beegle, E.J., Kavvlas, J.M. and G.T. Orlob, 1989. *Development of an expert system embedding pattern recognition technique for groundwater pollution source identification*, Completion Report, USGS Contract 14-08-0001-G1500.

David, R. and J. King, 1077. An overview of production systems, in *Machine Intelligence & machine Representations of Knowledge*, E. Elcock, and D. Michie (eds.), Wiley New York.

Deak, J., Deseo, E. and K. Davidesz, 1996. Verification of MODFLOW modelling in SE Hungary using environmental isotope and groundwater quality data, Proc. conf. *Hydroinformatics*, A. Muller (ed.), Zurich.

Deckers, F., 1994. A geohydrological information system based on object-oriented technology, Proc. conf. *Hydroinformatics*, A. Vervey, Minns, A., Babovic, V., and C. Maksimovic, (eds.), Balkema Rotterdam.

Doukidis, G.I., 1988. Decision Support System Concepts in Expert Systems: An Empirical Study, *Decision Support Systems* 4, pp. 345-354.

Dunn, S.M., Mackay, R., Adams, R. and D.R. Oglethorpe, 1996. The hydrological component of the NELUP decision-support system: an appraisal, *J of Hydrology (Special Issue: Decision-support systems)* Vol. 177 No. 3-4.

Eastern Research Group, 1993. *Subsurface characterisation and monitoring techniques, a desk reference guide, Part I*, EPA Cincinnati OH, pp. 491.

Eierman, M.A., Niederman, F. and C. Adams, 1995. DSS theory: A model of constructors and relationships, *Decision Support Systems* 14, pp. 1-26.

Evans, B.M. and W.L. Mayers, 1990. A GIS-base approach to evaluating regional groundwater pollution potential with DRASTIC, *J. Of Soil and Water Conservation*, vol. 45, no. 2, pp. 242-245.

Fayyad, Y., Piatetsky-Shapiro, G. and P. Smyth, 1996. From data mining to knowledge discovery in databases, *AI MAGAZINE* Fall issue.

Fedra, K. and H.J. Diersch, 1989. Interactive groundwater modelling: colour graphics, ICAD and AI, in *Groundwater Management: Quantity and Quality*, A. Sahuquillo (ed.), IAHS Publication 188.

Fedra, K., 1990. From useful to really usable: software for water resources planning and management, in *Transferring Models To Users*, E. Janes and R.W. Hotchkiss (eds.), AWRA Bethesda Maryland, pp. 73-86.

Fedra, K., 1993. Models, GIS, and expert systems: integrated water resources models, in *Application of GIS in Hydrology and Water Resources*, K. Kovar and H.P. Nachtnebel (eds.), IAHS Publication 211, pp. 297-309.

ForeFront Incorporated, 1994. *ForeHelp, a Help-authoring System for Microsoft Windows*, Version 1.0, ForeFront Incorporated, USA.

Furst, J., Girstmair, G. and H.P. Nachtnebel, 1993. Application of GIS in Decision Support Systems for Groundwater Management, in, *Application of GIS in Hydrology and Water Resources*, K. Kovar and H.P. Nachtnebel (eds.), IAHS Publication 211, pp. 13-22.

Gelhar, L.W., Welty, C. and K.R. Rehfeldt, 1992. A Critical Review of Data on Filed-Scale Dispersion in Aquifers, *Water Resources Research* 28(7), pp. 1955-1974.

GEOREF Systems Ltd., 1997. *Envirobrowser* (database) GEOREF Systems Ltd.

Goodall, A., 1997. *The Guide to Expert Systems*, Learned Information (Europe) Ltd. Abington.

Groen, J. and W.J. Zaadnoordijk, 1994. Selecting a number of optimal scenarios by means of genetic algorithm, Proc. conf. *Hydroinformatics*, A. Vervey, Minns, A., Babovic, V., and C. Maksimovic (eds.), Balkema Rotterdam 1994.

Grondin, G.H., Gannett, M., van der Heijde, P.K.M. and R.O. Patt, R.O, 1990. Critical errors that hydrogeologic professionals can make with computer programs, in *Transferring Models To Users*, E. Janes E., and R.W. Hotchkiss (eds.), AWRA Bethesda Maryland, pp. 149-158.

International Ground Water Modelling Centre, 1992. *MARS database summary reports*, IGWMC, Golden Co.

IWACO, 1995. *Active protection of drinking water resources in Hungary*, Interim report 1, IWACO Rotterdam.

Jamieson, D.G. and K. Fedra, 1996. The 'WaterWare' decision-support system for river-basin planning. 1. Conceptual design, *J. of Hydrology (Special Issue: Decision-support systems)* Vol. 177 No. 3-4.

Kinzelbach, W., 1986. *Groundwater modelling*, Elsevier.

Kolodner, J., 1993. *Case-Based Reasoning*, Morgan Koufmann.

Konikow, L.F. and J.D. Bredehoeft, 1992. Groundwater models cannot be validated, *Advances in Water Resources* 15, pp. 75-83.

Kukurić, N., 1998. Integration of models with REGIS/DINO products, TNO rapport NITG 98-248-B, the Netherlands.

Kukurić, N. and M.J. Hall, 1998. Electronic encapsulation of knowledge for groundwater quality management, *Water Resources Management* 12, pp. 51-79.

Kukurić, N., Hall, M.J. and Y. Zhou, 1998a. An Integrated Knowledge-Based Tool for Site Characterisation. Proc.conf. *Hydroinformatics '98*, Copenhagen, Denmark, Vol I, pp. 625-632.

Kukurić, N., Zhou, Y. and M.J. Hall, 1998b. An Integrated Knowledge-Based tool for Vulnerability Assessment. Proc.conf. *Hydrology in a changing environment*, Exeter, United Kingdom, Vol II, pp. 291-305.

Kukurić, N., Hall, M.J. and Y. Zhou, 1998c. Electronic Encapsulation Of The Knowledge For Groundwater Pollution Modelling, Proc.conf. *MODFLOW '98*, Golden, Colorado USA, pp. 621-628.

LeGrand, H.E., 1980. *A standardised system for evaluating waste-disposal sites*, National Water Well Association. pp.42.

Lopez, B., and E. Plaza, 1993, Case-based planning for medical diagnosis, in *Methodologies for Intelligent Systems*, J. Komorowski, and Z.W. Ras (eds.), Lecture notes in Artificial Intelligence 689, Springer-Verlag, pp. 96-105.

Loucks, P.D., Kindler, D.J. and K. Fedra, 1985. Interactive water resources modelling and model use: an overview, *Water Resources Research* 25(2), pp. 95-102.

Ludvigsen, P.J., Simsand, R.C. and W.J. Grain, 1986. A Demonstration Expert System to Aid in Assessing Groundwater Contamination Potential by Organic Chemics, in *Computing in Civil Engineering*, W.T. Lenocker (ed.), ASCE 1986, pp. 687-697.

LWI Werkgroep Generieke Tools, 1998. *Inventarisatie Generieke Tools*, Aquest rapport, CUR/LWI, The Netherlands.

McClymont, G.L., and F.W. Schwartz, 1987. *Development and application of an expert system in contaminant hydrogeology*, the expert ROKEY computer system, Final Report and Users Manuel, SIMCO Groundwater Research Ltd.

McClymont, G.L. and F.W. Schwartz, 1991a. Embedded knowledge in Software 1: Description of the system, *Ground Water*, vol. 29, no. 5.

McClymont, G.L. and F.W. Schwartz, 1991b. Embedded knowledge in Software 2: Demonstration and preliminary evaluation, *Ground Water*, vol. 29, no. 5.

McDonald, M. C. and A.W. Harbaugh, 1988. *MODFLOW, A modular three-dimensional finite difference ground-water flow model*, U. S. Geological Survey, Open-file report 83-875, Chapter A1, 1988.

McKinney, D. and Lin Min-Der, 1994. Genetic algorithm solution of groundwater management problems, *Water Resource Research*, 30(6), pp. 1897-1906.

McLaughlin, D. and W.K. Johnson, 1987. Comparison of three groundwater modelling studies, *J. Of Water Resources Planning and Management*, vol. 113 no. 3, pp. 405-421.

Mikroudis, G., 1988. GEOTOX manual, unpublished.

Minsky, M., 1975. A framework for representing knowledge, in *The psychology of computer Vision*, P. Winston (ed.), McGraw-Hill.

National Research Council, 1990. *Ground Water Models: Scientific and Regulatory Applications*, National Academy Press, pp. 303.

National Research Council, 1993. *Ground Water Vulnerability Assessment*, National Academy Press 1993, pp. 204.

Newell, C.J., Hopkins, L.P. and P.B. Bedient, 1990a. A Hydrogeologic Database for Ground-Water Modelling, *Ground Water*, vol. 28, no. 5, pp. 703-714.

Newell, C.J., HaasBeek, J.F. and P.B. Bedient, 1990b. OASIS: A Graphical Decision Support System for Ground-Water Contamination Modelling, *Ground Water*, vol. 28, no. 2, pp. 224-234.

Olsthoorn, T.N., 1995. Effective Parameter Optimisation for Ground-Water Model Calibration, *Ground Water* Vol 28 no 2, pp. 42-48.

Olsthoorn, T.N. (ed), 1996. *Model calibration* (Modelcalibratie), conference proceedings, Nederlandse Hydrologishe Vereniging (in Dutch).

OSWER Information Management, 1994. *Assessment Framework for groundwater model application*, US EPA 500-B-94-003, pp. 41.

Padget, D.A., 1994.Using 'DRASTIC' to improve the integrity of GIS data used for solid waste management   facility siting: a case study, *The Environmental Professional*, vol. 16, no. 3, pp. 211-219.

Peck, A., Gorelick, S., De Marsily, G., Foster, S. and V. Kovalevsky, 1988. *Consequences of spatial variability in aquifer properties and data limitations for groundwater modelling practice*, IAHS Publication 175.

Porter, B.W., Bareiss, R., and R.C. Holte, 1990. Concept learning and heuristic classification in weak-theory   domains, *Artificial Intelligence* 45, pp.229-263.

Price, R.K., 1997. *Hydroinformatics, Society and the Market*, inaugural address, IHE The Netherlands.

Quercia, F., 1993. *Summary review of available models for groundwater flow and contaminant migration*, UNESCO Paris, pp. 58.

Radermacher, F.J., 1994. Decision support systems: scope and potential, *Decision Support Systems* 12, pp. 257-265.

Rail, C.D., 1989. *Groundwater Contamination (Sources, Control and Preventive Measures)*, Technomic Publishing.

Reddi, L.M., 1990. Potential pitfalls in using groundwater models, in *Transferring Models To Users*, E. Janes E., and R.W. Hotchkiss (eds.), AWRA Bethesda Maryland, pp. 131-140.

Riesbeck, C.K. and R.S. Schank, 1989. *Inside case-based reasoning*, Northvale NJ Erlbaum.

Rifai, H.S., Hendricks, L.A., Kilborn, K. and P.B. Bedient, 1993. A Geographic Information System (GIS) User Interface for Delineating Wellhead Protection Areas, *Ground Water*, vol. 31, no. 3, pp. 480-488.

Rifai, H.S., Bedient, P.B. and C.J. Newell, 1994. Decision support system for evaluating pump-and-treat remediation alternatives, *Computer Techniques in Environmental studies 5*, vol. 1, MA USA, pp 219-226.

Ritzel, B.J. and J.W. Eheart, 1994. Using genetic algorithm to solve a multiple objective groundwater pollution containment problem, *Water Resources Research*, 30(5), pp. 1589-1603.

Roaza, H., Roaza, R.M., and J.R. Wagner, 1993. Integrating Geographic Information Systems in Ground-Water Applications Using Numerical Modelling Techniques, *Water Resources Bulletin* vol. 29, no. 6, pp. 981-988.

Rouhani S., and R. Kangari,1987. Expert system in water resources, in *Water for the future: Hydrology in perspective*, J. Rodda and N.C. Matalas (eds.), IAHS Publication 164, pp. 457-462.

Rumbaugh, J., Blaga, M, Premerlani, W., Eddy, F. and W. Lorensen, 1991. *Object-oriented modelling and design*, Prentice-Hall.

Rundquist, C.D., Rodokohr, D.A., Peters, A.J., Ehrman, R.L., Di, L. and G. Murray,1991. Statewide Groundwater Vulnerability Assessment in Nebraska Using the DRASTIC/GIS Model, *Geocarto International* (2).

Russel, S. and P. Norvig, 1995. *Artificial Intelligence, a modern approach*, Prentice Hall, pp. 930.

Saaltink, M.W. and J. Carrera, 1992. *A Knowledge based System for Modelling Soil and Groundwater Pollution*, TNO report OS-92/113A.

Schank, R. and R. Abelson, 1977. *Scripts, Plans, Goals and Understanding*, Northvale NJ Erlbaum.

Schank, R., 1982. *A theory of learning in computers and people*, Cambridge Univ. Press.

Scientific Software Group, 1995. *MODEL EXPERT 95*, Scientific Software Group.

Sims, R.C., Sims, J.L., and S.G. Hansen, 1991. *Soil Transport and Fate Database and Model Management System*, R.S. Kerr Lab. US EPA.

Slade, S., 1991. Case-Based Reasoning: A Research Paradigm, *AI Magazine*, pp. 42-55.

Teso, R.R., Younglove, T., Peterson, M.R., Sheeks, D.L. and R.E. Gallavan, 1988. Soil taxonomy and surveys: Classification of areal sensitivity to pesticide contamination of groundwater. *Journal of Soil and Water Conservation*, vol 43. no. 4.

Sokol, G., 1996. Development of a GIS-based user shell for hydrogeological applications, in *Application of GISs in Hydrology and Water Resources Management*, K. Kovar and H.P. Nachtnebel (eds.), IAHS Publ. no. 235.

Sotornikova, R. and J. Vrba, 1987. Some remarks on the concept of vulnerability maps, in *Vulnerability of soil and groundwater to pollution*, W. van Duijvenboden and H.G. van Waegeningh (eds.), TNO Committee on Hydrological Research, The Hague, Proceedings and Information no.38, p471-476.

StreamLine Groundwater Applications, 1996. *Single Well Solution Software*, StreamLine Groundwater Applications, USA.

TNO, 1994. *Regional Geohydrological Information System [REGIS]*, TNO Institute of Applied Science, The Netherlands.

UNESCO, 1995. *GWW Software version 1.1*, United Nations, New York.

US. EPA, 1992. *The Hazard Ranking System guidance manual*. U.S. EPA Office of Emergency and Remedial Response, Washington DC. Pp. 382.

Van Duijvenboden, W. and van H.G. Waegeningh (eds.), 1987. *Vulnerability of soil and groundwater to pollutants*, TNO Committee on Hydrological Research, The Hague, Proceedings and Information No.38.

Van der Heijde, P.K.M., Bachmat, Y., Bredehoeft, J., Andrews, B., Holtz, D. and S. Sebastian, 1985. *Ground-Water Management: The use of numerical models*, Water Resource Monograph 5, AGU Washington D.C.

Van der Heijde, P.K.M. and A.O. Elnawawy, 1993. *Compilation of Ground-Water Models*, IGWMC, Colorado School of Mines, p.289, Golden Co.

Villumsen, A., Jacobsen, O.S., and C. Sonderskov, 1982. Mapping the vulnerability of groundwater reservoirs with regard to surface pollution *Geological Survey of Denmark*, Yearbook 1982, Copenhagen, pp. 17-38.

Vrba, J. and A. Zaporozec, 1994. *Guidebook on Mapping Groundwater Vulnerability*, vol. 16, IAH, pp. 131.

Waterman, D.A., 1986. *A Guide to Expert Systems*, Addison-Wesley Publishing Co.

Walsh, M.R., 1993. Towards spatial support systems in water resources, *J. of Water Resources Planning and Management*, vol.119, no. 2, pp. 158-169.

Weiss, S.M., and C.A. Kulikowski, 1984. *A Practical Guide to Designing Expert Systems*, Rowman and Allanheld Publ. New York.

Wielinga, B., Schreiber, G. and J. Breuker, 1993. Modelling Expertise, *KADS, a principled approach to Knowledge Based system development*, Schreiber, G. Wielinga, B. and J. Breuker (Eds), Academic Press.

Wilson, J.L., Mikroudis, K. and H.Y. Fang, 1987. GEOTOX: A knowledge-based system for hazardous site evaluation, *Artificial Intelligence* 2(1) 1987, pp. 662-667.

Yeh, W.W., 1986. Review of parameter identification procedures in groundwater hydrology: The inverse problem, *Water Resources Research* 22(2) pp. 95-108.

Zomorodi, K., 1990. Experiences in using Modflow on a PC, in *Transferring Models To Users*, E. Janes E., and R.W. Hotchkiss (eds.), AWRA Bethesda Maryland, pp. 351-355.

ANNEX: BEYOND ENGINEERING

It is all about emotion
Logica

The DSS is a software tool, therefore the research presented in this thesis was about software engineering. It was also about knowledge engineering, because the DSS contains knowledge. Engineering is, however, not sufficient to comprehend contemporary software development; this especially holds for electronic encapsulation of knowledge. It is necessary to establish one's own concept (or vision) of Information Technology, whatever that might be.

Several years ago, Internet was not much more than a pile of advertisements. Nowadays, it is a tool used on daily basis by millions of people all over the world. Internet is just a good example of what has happened with IT (or what IT does to us) in a short period of time; new information technologies are not simply tools to be applied, but processes to be developed. Users and doers may become the same. The users can take control of the technology (the opposite will not be discussed here), as in the case of Internet.

Advances in IT are so rapid and influential that a new, Information Technology revolution is proclaimed. Moreover, the IT revolution is recognised as the 'third revolution' (coming after those of agriculture and industry), and informationalism as a new mode of development. Through IT, a close relationship has been established between processes of creating and manipulating symbols and the capacity to produce and distribute goods and services. The human mind is nowadays, for the first time in history, a direct productive force, not just a decisive element of the production system.[1] Gradually, computers (and computer/communication systems) are becoming amplifiers and extensions of the human mind. In these terms, an *initial impulse* (thought, idea, data, knowledge) comes, in principle, always from the mind; additionally, the impulse should be understandable to a computer. Apparently, informationalism is conditioned by:

– a willingness of the mind (person) to communicate 'with computer', i.e. share the thoughts (ideas, data, knowledge) with others via a communication system, and
– (an appropriate) mode of communication.

The former issue is primarily a matter of power and ethics, and will be discussed later on. The latter is subject not only of software (knowledge) engineering, but also of AI.

---

[1] The relation between various modes of production (capitalism, statism) and modes of development (e.g. industrialism) has changed with the emergence of informationalism. 'What is specific to the informational mode of development is the action of knowledge upon knowledge itself as the main source of productivity' (Castells, 1996).

'Feeding' the computer with numerical information has become routine; numerous databases are available to store various kinds of data. Processing of numerical information was (and still is?) the computer's 'core business'. Various possibilities for visualisation of (processed) data are nowadays the most impressive computer feature, but what can be said about electronic storage, processing and visualisation of knowledge?

What is knowledge? A partitioning of information on quantitative (numerical) and qualitative information (knowledge), as made in this research, is essentially relative, because in a broader sense information as a whole (i.e. qualitative and quantitative) can be seen as knowledge.[2] This partition is, however, made to highlight knowledge that is not (or cannot be) expressed in digits, and processed and presented as such (see Chapter 2 and Chapter 3).

Because of the dominant role of modelling, numerical information still attracts most attention; modelling is so influential that some consider modelling as knowledge processing; they are not wrong, but there is only a portion of human knowledge that could (and should) be subjected to this kind of processing. Some descriptive (non-numerical) information can be stored in contemporary databases and subsequently quantified (represented by numerical values); this allows (numerical) processing of knowledge, as it is done, for example, by rating techniques (Chapter 5). Various techniques developed in the field of AI (including very popular 'self learning' techniques, such as neural networks and genetic algorithms) can be used in the same manner for 'knowledge processing'. These techniques are meant for numerical processing - or to be more precise - modelling. Within hydroinformatics, their application is often referred to as (sub-symbolic) metamodelling.[3] In his characterisation of hydroinformatics Abbott (1991) describes (by quoting Heidegger) information that 'enters into hydroinformatics':'...it is to say, as a first approximation, that it is information that is orderable, countable and subject to computation'. That certainly holds for numerical information; knowledge also needs to be 'ordered' to provide/improve communication, but it is not necessary countable and subject to computation.[4]

Most likely, mathematics emerged from attempts to quantify human thoughts and improve communication. One can recall Descartes (a founder of modern physics) and his Cartesians, who

---

[2] There are numerous definitions of knowledge, some of which are given in this and following chapters. In particular relation between knowledge and information is understood differently by different people. For many, information is simply communication of knowledge. According to Abbott (1991), 'knowledge per se cannot be replaced by information'; '...above information is knowledge and above knowledge is wisdom'.

[3] 'Hydroinformatics is (appropriate) information technology for (modelling and) managing water' (Price, 1997). Within 'managing' Price gathered a number of activities (acquisition, analysis, communication, etc.), including 'decision support'. Nevertheless, the main role is still played by modelling.

[4] 'Knowledge: a set of *organized* statements of facts and ideas, presenting a reasoned judgement or an experimental result, which is transmitted to others through some communication medium in some *systematic* form' (Bell, 1973).

tried to set out their philosophy by using geometric reasoning. Similarly, Spinoza took the geometry of Euclid as his model of objective rational enquiry. Without challenging strong connection between logic and mathematics, between semantic and semiotics (see Section 2.4.3), cognition is (still) too 'fuzzy' to be completely quantified (described and modelled mathematically).

AI, in its most genuine form, deals with human thinking, trying, eventually, to formalise and encapsulate knowledge which emerges therefrom (see Section 2.4). This very difficult but challenging task involves social, philosophical, ethical, and a number of other interwoven facets. In that context, statements like 'does thinking really involve thinking?' (Riesbeck and Schank, 1989) sound very logical to people who struggle with electronic encapsulation of knowledge.[5]

During the last few decades AI has attempted to formalise (and encode) human thoughts in sets of rules (and more complex rule-based forms). These attempts have been fairly successful for well-structured, narrow knowledge domains. However, the 'classical' rule-based approach fails when cognition is about case-specific problems (no generalisation possible) and/or interdisciplinary problems (where considerable knowledge from various field is required). The AI task becomes even more difficult when knowledge is incomplete and/or inconsistent (which is often the case). Various approaches have been suggested to deal with these problems during the encapsulation process, as well as once knowledge is encapsulated (case-based reasoning, inconsistency-tolerant logics, knowledge discovery techniques, etc.). Rapid enlargement of encapsulated knowledge has, in particular, urged development of techniques that would assist (re)structuring and more efficient (re)use of already encapsulated knowledge. This acting upon already encapsulated knowledge is nowadays often called *knowledge processing* - see Section 3.5. Success of both knowledge encapsulation and knowledge processing depends heavily on communication ('willingness' and 'mode' see p. 4).

Knowledge empowers; some say 'knowledge is power'.[6] By communicating people transfer the power. Besides, communication is often a *conditio sine qua non* for creation of a new piece of knowledge (e.g teamwork). Quite recently, IT changed its name into Information and Communication Technology (ICT). Most of newly-developed ICT techniques (e.g. groupware technology, Intranet/Extranet technology) are meant to support and/or promote electronic communication, cooperation and coordination. A completely new *network society* emerges; 'the power of flows takes precedence over the flow of power' (Castells, 1996). Nation states, as we know them, are founded upon production, knowledge (about self and others) and power. The new mode of development (informationalism) frees production from a place or territory;

---

[5] Even hydroinformatics, which claims to be 'only' technology, has its challenges. In his account Abbott (1991) stated that '...philosophical level is not enough;... only at theological level we can establish a proper conceptual foundation for hydroinformatics'.

[6] Power is seen here as relation between human subjects which, on the basis of *production* and *knowledge*, imposes the will of some subjects upon others by potential or actual use of violence, physical or symbolic.

communication becomes power which is difficult to control by a state (the same holds for the context of communication-knowledge).

An interesting example is capital flow in the world (relation between knowledge and capital, i.e. information and investments, needs no explanation). Electronic transactions accounted for £150 billion of the global economy in 1996. By the year 2005 it could be as high as £2 trillion. Ian Angell (professor at London School of Economics and one of the gurus of the information age) believes that the liquidity of the electronic cash will bring about the collapse of the tax system and precipitate the end of the nation state. 'Companies will use the Internet to disappear into cyberspace, and because it will then be impossible to derive revenue from them, governments will be forced to make drastic changes to maintain the status quo' (The TIMES, July 9, 1997). And further: 'The new elite will be skilled "knowledge workers", mercenary in their choice of places to live, connected via a network with the full panoply of teleworking technology at their command. This will lead to an unprecedented fragmentation of the countries and nation states which we take for granted today.'

The importance of communication is more than obvious; without proper communication and cooperation knowledge encapsulation and processing remain far below the optimal. A 'team spirit' is requested in vast majority of job advertisements, ICT techniques are developed to provide or enhance 'collaborative decision making', and yet, something is missing. When related to power (see previous footnote on capital flow) communication appears to be excellent. Does everything have to be seen through power in order to function, in order to be worth doing? Is power the only driving force of mankind?[7]

Knowledge and ethics (personal note)

There is certainly another driving force which counteracts power and it will be addressed at the end of this section. Power and related concepts, however, ask for more attention.

Power is founded upon human characteristics such as egoism, fear and hypocrisy. 'Status and rewards go to knowledge owners' is recognised as one of the major obstacles to 'knowledge transfer' in one of recent publication of Harvard Business School. Why should one share knowledge in this materialistic world if that act is not rewarded through (further) empowerment? Egoism is natural and congenital (defect?), and as we grow up, we learn to deal with it (to limit or suppress it). That is done in variety of ways, but because of just one reason: we cannot live alone. '...and you would most likely die or become crazy if you would lived in a society where everyone had decided not to look at you and to behave like you did not exist' wrote Umberto Eco in his 'Five moral dilemmas'. Bitterly laughing at egoism, La Bruyère said: 'all our evil originate from the fact that we are not able to be alone'. In the same line, humanistic psychology

---

[7] Additional note on capital flow. James Tobin, Nobel Prize winner, suggested a long time ago (1972) the tax of 0.5% to be imposed on capital flow in the world. He was supported by ex-president of France Mitterrand and some left-wing economists. The same suggestion was recently repeated in the Dutch parliament (and a 'liberal' minister of economy was not amused). The suggestion was obviously made in an attempt to control and limit power. Why?

acknowledges a perception of 'belonging to' as one of decisive for human well-being (e.g. Maslow in 'Towards a Psychology of Being').

Consequently, people tend to group around primary identities (religious, ethnic, territorial) building up power-based relations with other groups. Grouping provides security by enlargement of power; grouping is, however, also a search for identity (collective or individual). This search has been present since the dawn of human civilisation, but it becomes increasingly important in a world of global flows, a world full of uncontrolled, confusing changes. For many, the search for individuality is, unfortunately, nothing more than conformism - escape from separation and loneliness. 'They live in illusion that they follow their own ideas and intention, that they form opinions individually, and it is just a matter of pure coincidence that their ideas coincide with those of majority' (Fromm). Anyway, a repercussion is an 'individualistic', self-sufficient and egocentric behaviour.[8]

Then the fear; my experience, my expertise, my heuristic knowledge is my wealth and my sword, and if I share it 'just like that', I will become poor and defenceless.[9] Well, if Voltaire was right, and if *homo homini lupus (est)*, no objection could be raised up against these thoughts. Objectively, however, knowledge sharing is the best way for knowledge enlargement. Besides, if the other driving force (and not power) is exercised, unselfish knowledge sharing is simply an imperative.

There is an additional source of fear; lack of knowledge and/or insecurity in one's own knowledge create the fear of being 'uncovered'. Sometimes it is difficult to avoid the impression that all that ado about 'expert knowledge' is created by 'experts' themselves as an attempt to esoterically defend their position. Anyway, it is almost a rule: the more knowledge one has, the more one is prepared to share it.

Finally, hypocrisy - maybe not the darkest, but certainly the most nauseating human characteristic. 'Alike our body wrapped in clothes, our mind is wrapped in lies. Our communication, our work, our complete being is false; and only through that shell a sight of our genuine character can be caught, like the sight of a body can be caught through the clothes' (Schopenhauer).

---

[8] ' As time goes on, people always want more: more fame, more to possess, higher, bigger, better. They cannot stop. Nowadays we read a lot about interest rates (in stock-markets). Would you like, as a future investor, the highest possible interest rate? Then invest once in your fellow human being.' (Anonymous, Delftse Post, January 3, 1997)

[9] 'In this world, one can succeed only with tip of his sword, and dies with a weapon in his hand' (Voltaire).

'In life, public opinion, conflict of interest and ambition force a man to hide his weaknesses, to cover up shortcomings and patches, and not to reveal his most secret thoughts to the wide world. And the best outcome of that unusually necessary practice is that you, by deceiving the others, finally deceive yourself, because in that case you are spared of shame, which is very unpleasant feeling, and of hypocrisy, which is a disgusting characteristic. And when you are dead, what a difference! What a relief, what a freedom! Man can, without hesitation, to throw away the mask, to bang all that false jewellery on a floor, to thaw, to change appearance, cloths, to confess everything, to say what was he really alike, and what was he pretended to be alike. No more neighbours, friends, enemies, acquaintances and strangers, no consideration towards external world - there is nothing left! Even public opinion disappeared.... And to your knowledge, gentlemen alive: nothing can be compared with dead-man's despise'.

(Machado de Assis in 'Memorias posthumas de Bras Cubas')

'O tempora! O mores!' Someone spoke out (shouted) these timeless words even before Christ was born. Hypocrisy is very much related to egoism and fear, being a major obstacle in the search for knowledge, the search for truth. 'So far in me lies, I value, above all other things out of my control, the joining hands of friendship with men who are lovers of truth' (Spinoza).

Is the truth (the search for truth) that other driving force? Or is it love?[10] Does one have to believe in altruism or in God ('many are called, a few chosen') in order to exercise this 'positive approach'?[11] Or is everything just a struggle between good and evil (in us and around us)? [12, 13]

Work with knowledge (and some other factors) brought along these questions (and lot of dilemmas as well). They are mentioned here as being crucial for optimal encapsulation and processing of knowledge. They are also decisive for a number of other issues; those are, however, not a subject of this thesis.

It will certainly not be very original to close this introduction by quoting Augustine; unfortunately, I do not see a better way:

---

[10] 'Love is primarily "giving",...care, responsibility, respect and knowledge are common elements for all kinds of love' (Fromm)

[11] 'Jesus was a street-corner philosopher whose primary question was: 'How can I become a good person in a bad world?' (B. Miller in Elsevier, December 19, 1998)

[12] And what is a role of 'conscience' in that struggle; 'is conscience nothing more than the suggestive working of prescribed moral norms, or *vox Dei*?' (Jung in 'Good and Evil').

[13] In many discourses on good and evil (like following of Spinoza) the heaven is illuminated, but stairways are not more than a vague imagination. According to Spinoza, in our ordinary lives, when seeing the world *sub specie durationalis*, we see also the great divide between good and evil, and we are torn between the demands of morality and the 'temptations of nature'. Those attitudes, however, derive from the first (lowest) level of cognition (knowledge gained through sense-perception is assigned to this level). In short, good is not defined by this or that person's moral judgement, but objectively, by human nature. Good is what we certainly know that is useful (for us); but we can know what is useful only if we learn to see things from the impartial point of view of the rational observer to whom things appear 'under of aspect of eternity' (*sub specie aeternitatis*).

Pondus meum amor meus, eo feror quocumque feror (my love is my weight; by it I am carried wherever I am carried).

dixi

SAMENVATTING

Computers zijn het belangrijkste gereedschap van het grondwaterbeheer geworden. Computer software is ontwikkeld voor de opslag, verwerking en presentatie van informatie over de verontreinigingsproblematiek van grondwater. De aanhoudende behoefte voor een efficiëntere informatiebehandeling heeft tot een toenemende integratie van software in zogenaamde Decision Support Systemen (DSS-en) geleid. Zowel kwantitatieve (numerieke) als kwalitatieve (kennis) informatie is vereist om besluitvormingsprocessen te ondersteunen. Een uitvoerig overzicht van de beschikbare software (dit proefschrift en publicatie in *Water Resources Management*) heeft aangetoond dat tot nu toe de ontwikkeling en de integratie van software veel meer gewijd is geweest aan numerieke informatie dan aan kennis. Zonder een kenniscomponent blijft het DSS een geïntegreerd softwaresysteem, terwijl het voornamelijk een *geïntegreerde informatiesysteem* zou moeten zijn. Aangezien informatie een cruciale rol in besluitvormingsprocessen speelt, zou het als het belangrijkste criterium tijdens het bepalen van de opbouw van een DSS genomen moeten worden. In principe is de steun die een DSS kan verstrekken proportioneel aan de (kwalitatieve en kwantitatieve) informatie die het DSS bevat.

In deze thesis beschreven het 'DSS for Groundwater Pollution Assessment' is ontworpen als een geïntegreerd informatiesysteem. Het systeem is opgebouwd uit een DSS kern, een applicatieomgeving en een kenniscomponent. Het ontwerp en de ontwikkeling (en integratie) van de kenniscomponent zijn de stappen van het *knowledge encapsulation process*. Dit process is veel meer dan een transformatie van kennis in elektronische vorm en het opslaan van kennis in een computer. Het 'inkapselen' begint met het definiëren van het probleem en omvat ook de net zo belangrijke acquisitie, systematisering en formalisering van de kennis. De 'probleem-specifieke' en interdisciplinaire karakter van grondwaterproblematiek is de belangrijkste obstakels voor een effectieve kennis 'inkapseling'. Vanwege deze en andere obstakels is de kennis die ingekapseld wordt vaak incompleet, inconsistent, slecht gestructureerd en gebrekkig aan geformaliseerde expertise.

De kenniscomponent van het 'DSS for Groundwater Pollution Assessment' is samengesteld uit drie Knowledge-Based Modules (KBMs). De inhoud en het ontwerp van de KBMs zijn sterk afhankelijk van de taken die zij ondersteunen. Echter, in principe verstrekken zij informatie over de acties die voor de uitvoering van de taak nodig zijn; welke operatie vereist of aanbevolen zijn (bv. acquisitie, verwerking, presentatie, enz.), waarom, hoe ze uit te voeren en wanneer. Tot nu toe zijn drie KBMs ontworpen en ontwikkeld, te weten: de Site Characterisation Module (SCM), de Vulnerability Assessment Module (VAM) en de Groundwater Modelling Module (GMM).

Het karakteriseren van het onderzoeksgebied is de eerste taak bij de aanpak van een grondwater verontreinigingsprobleem. De voornaamste doelen van de karakterisering zijn de conceptualisering van het grondwatersysteem en de diagnoses van de grondwaterverontreiniging. Algemene kennis over de karakterisering van een onderzoeksgebied is verzameld, gesystematiseerd en ingekapseld in de Site Characterisation Module. De SCM is in de HyperText Markup Language (HTML) gecodeerd, een taal die het mogelijk maakt grote hoeveelheden van de semi-gestructureerde kennis in te kapselen in een gemakkelijk aan te passen vorm. Naast kennis, bevat de SCM ook de verbindingen met procedures voor de opslag, verwerking en presentatie van data binnen het DSS. De module is gepresenteerd tijdens de internationale conferentie 'Hydroinformatics 98' (Kopenhagen, Denemarken) en in het conferentieverslag gepubliceerd.

'Vulnerability Assessment' (VA) van grondwater kan in algemene termen beschreven worden als een procedure voor een snelle inschatting van de grondwaterverontreiniging. De procedure wordt op intrinsieke kenmerken van de aquifer gebaseerd, hoewel ook de kenmerken van de schadelijke stof en het waterbeheer in beschouwing genomen kunnen worden. In het kader van onderzoek naar de grondwaterverontreiniging (op lokale schaal), volgt de VA-stap na de karakterisering van het onderzoeksgebied. De VA gebruikt het resultaat van de karakterisering om de eerste inschatting van de verontreinigingspotentiaal te bepalen. Deze inschatting dient als basis voor verder onderzoek en/of ter vergelijking van verontreinigingspotentialen. Een VA rangordemethode is ontwikkeld en in de Vulnerability Assessment Module ingekapseld. De VAM is ontwikkeld met gebruik van het object-georiënteerde Delphi Developer gereedschap. Bij integratie van de VAM in het DSS zijn de DSS kern, de SCM en een Chemische Database (CB) gelinkt. De module is gepresenteerd tijdens de internationale conferentie 'Hydrology in a Changing Environment' (Exeter, Verenigde Koninkrijk) en in het conferentieverslag gepubliceerd.

De 'Groundwater Modelling Module' (GMM) is ontwikkeld met de bedoeling om het modelleren van verontreinigingsproblemen op lokale schaal te ondersteunen. De kern van de GMM is een elektronische Modelling Protocol handleiding, samengesteld uit een set van hypertext-gebaseerde topics. De kennis over algemene complexiteit van het model is in een rule-based 'Model Complexity Module' ingekapseld. Dat is gedaan met gebruik van een knowledge-base editor. De GMM bevat ook (in Delphi ontwikkelde) modulen die assisteren bij inschatting van de retardatiefactor en de dispersiecoëfficiënt. Het Modelling Protocol dient als een platform dat software applicaties in een unieke DSS module integreert. Als onderdeel van de GMM ontwikkeling is MODFLOW, een modulair 3D grondwaterstroming model, volledig met de DSS kern geïntegreerd. De module is gepresenteerd tijdens de internationale conferentie 'MODFLOW '98' (Golden, Colorado) en in het conferentieverslag gepubliceerd.

Verschillende vormen voor de representatie van kennis aangeboden door Kunstmatige Intelligentie zijn gebruikt bij de ontwikkeling van de KBMs. Vanwege een tamelijk slechte taxonomie van kennis over de grondwater-verontreinigingsproblematiek, kon maar een klein deel van de kennis in regels uitgedrukt worden. Echter, de hypertext-based technologie bewees zich als geschikt voor het inkapselen van grote hoeveelheden semi-gestructureerde kennis. De nieuwe 'mark-up' talen (bv. XML- eXtended Mark-up Language) die recentelijk op de markt geïntroduceerd zijn, verstrekken veel betere mogelijkheden voor inkapselen dan de 'klassieke' HTML (gelukkig de conversie is mogelijk). Deze talen, gecombineerd met Java applets (Java neemt het programmeringsveld over van andere talen), creëren hedendaagse Intranet/Extranet werelden.

Case-Based Reasoning (CBR) is een veelbelovende aanpak voor elektronische inkapseling en verwerking van kennis. Voorstanders van CBR hebben de hypothese verworpen dat menselijke gedachten van een set van redeneerprincipes of regels afhankelijk zijn. Zij hebben als primaire kennisstructuur zogeheten 'scripts', veel grotere hoeveelheden van kennis, geïntroduceerd. Hun hypothese luidt dat men een gebeurtenis onthoudt in termen van geassocieerde scripts, d.w.z. cases. Het idee achter CBR is om de cases in een kennisbank op te slaan en hen te gebruiken als men een nieuw probleem tegenkomt. De cases die lijken op het nieuwe probleem dienen vanuit de kennisbank geselecteerd te worden en de 'oude' oplossingen dienen aangepast te worden om het nieuwe probleem op te lossen. Conform daarmee dient CBR de volgende taken uit te voeren: de actuele probleemsituatie identificeren, een oude case die vergelijkbaar is met het nieuwe vinden, de gekozen case gebruiken om een oplossing voor het nieuwe probleem voor te stellen, voorgestelde oplossing evalueren en de kennisbank aanvullen door de nieuwe ervaring. Deze dissertatie bevat een uitgebreide beschrijving van CBR postulaten. Daarnaast is aan de voorwaarden voor toepassing van CBR bij grondwatermodellering voldaan door de ontwikkeling van de GMM. De belangrijkste voorwaarden zijn kwaliteit en consistentie van de gemodelleerde cases. Een omvangrijk aan CBR gewijd onderzoeksproject is recentelijk begonnen bij NITG-TNO.

De razendsnelle opkomst van Informatie en Communicatie Technologie (ICT) is nieuwe grenzen aan het openen voor de inkapseling van kennis en voor DSS ontwikkeling in het algemeen. Dankzij zijn flexibiliteit en aanpassingsvermogen wordt hedendaagse software 'interactief' in de echte betekenis van de term; de kennis wordt naar de gebruiker overgebracht, terwijl de gebruiker een mogelijkheid krijgt om zijn eigen kennis aan de computer te geven en daarmee de software te verbeteren. Nieuwe informatie technologieën zijn niet alleen een gereedschap om toe te passen, maar ook processen om te ontwikkelen. Gebruikers en ontwikkelaars kunnen hetzelfde worden. Dit wordt zelfs belangrijker in de combinatie met Internet en Internet-gerelateerde technieken; zij maken ingekapselde kennis toegankelijk voor een buitengewoon brede cirkel van potentiële gebruikers/ontwikkelaars. ICT technieken zouden toegepast moeten worden om individuele samenwerking tussen experts, maar ook die tussen verschillende velden van expertise (bv. hydrogeologie, geochemie, ecologie, software engineering, enz.) te verbeteren. Een oprechte samenwerking (een echt teamwork) is evenzeer van belang voor het vaststellen van de fundamentele taxonomie van grondwaterbeheer als voor software integratie; en voor alles ertussen.

CURRICULUM VITAE

Nebojša (Neno) Kukurić graduated from the University of Belgrade in Hydrogeology in 1986, and joined The Karst Water Research Institute (KWRI) in Trebinje (Yugoslavia). For the next five years, he was involved in planning and supervising all kinds of hydrogeological exploration as a part of the complex project Trebisnjica HydroSystem - second stage of development. At the same time N. Kukurić continued post-graduated education at University of Belgrade, obtaining MSc in Hydrogeology and, subsequently, conducting the PhD research. The research and the work at KWRI were abruptly terminated in 1991 by the war that broke out in former Yugoslavia.

In 1992 N. Kukurić completed a year long post-graduate course in Hydrological Engineering at IHE Delft, followed by MSc degree obtained a year later (all academic degrees were obtained with distinction). From 1992 to 1995 N. Kukurić worked as hydrogeologist at IWACO, gaining wide experience in groundwater modelling, multi-criteria analysis, geostatistics and time-series analysis. In 1995 he returned to the IHE to carry out his doctoral research dedicated to electronic encapsulation of knowledge in groundwater management. Since 1998, Neno Kukurić is a staff member of the Netherlands Institute of Applied Geoscience TNO.